Some chapters of *A BEAUTIFUL IVORY BANGLE* were included in *Seashore Man & African Eve,* published in 2007. Substantial revisions resulted in a second edition of *Seashore Man & African Eve.* Those portions which were not precisely relevant to the theme of evolution of mankind were edited out and it was decided to rework them into a new book.

A Beautiful Ivory Bangle explores changes in sub-Saharan African societies which may have been caused by the activities of the Indian Ocean trading system, initiated from northern hemisphere civilisations, during the last 5,000 years, leading to the exploration of the Indian Ocean by the Portuguese in the late 15th century.

Denis Montgomery was born in 1934 in Natal, South Africa, and during his varied career he travelled widely in Africa and around the Indian Ocean rim. He lived and worked in Nigeria, South Africa, Mozambique, Brazil and England but whenever possible he travelled. It was tourism projects in the wilderness that spurred his early fascination with African pre-history and modern population and environmental impacts. Since the 1960s, these have been his abiding passions. He was elected a Fellow of The Royal Geographical Society in 1989.

Cover picture : A replica Omani *boum* off Zanzibar in 2006
(Author's photo)

A Beautiful Ivory Bangle

Eastern Africa and Seatraders

Denis Montgomery

What's past is prologue
Shakespeare

Other publications:

The Reflected Face of Africa. 1988
            second edition 2006
Two Shores of the Ocean. 1992
Crest of a Wave. 2006
Seashore Man and African Eve. 2007
            second edition 2008
Mud, Sands & Seas. 2007

© Denis Montgomery 2008
This book is copyright. All rights are reserved.

Produced with WordPerfect v 14.0 in Palatino Linotype
Photographs from the author's collection.

African Insight, 2008
41 Majors Close
Chedburgh
Suffolk, England
www.sondela.co.uk

Published by Lulu in the United States of America
www.lulu.com
Lulu Ref. 3407283

ISBN 978-1-4357-5953-4

Contents

Illustrations, quotations and notes

Introduction 19

ONE : Northern Hemisphere Classical Civilisations 23
    Agriculture, trading and Civilisation

TWO : A Vortex in East Africa 49
    Eastern central Africa, an entrepôt for cultures

THREE : The Khoisan 71
    The mysterious spread of nomadic herding to the far south

FOUR : Cattle point the way 117
    Tracing the introduction of modern cattle to Africa

FIVE : Indian Ocean Seatraders 127
    The great Indian Ocean trading system

SIX : The Swahili Coast 153
    The impact of seatraders on eastern Africa.

SEVEN : A beautiful ivory bangle 191
    The rapid colonisation by Bantu-speaking mixed agriculturalists.

EIGHT : The Golden Rhino and Zimbabwe 221
    Indian Ocean trading influences in southern Africa.

NINE : *Terra da Boa Gente* 247
    Western Europeans explore south of the Sahara.

Bibliography & further reading 279

General Index 289

Illustrations are from the author's collection.

Author's photo

A faithful replica of an Omani *boum* off Zanzibar in 2006

Quinquereme of Nineveh from distant Ophir,
Rowing home to haven in sunny Palestine,
With a cargo of ivory,
And apes and peacocks,
Sandalwood, cedarwood, and sweet white wine.

From *Cargoes* by John Masefield

SUNDRY QUOTATIONS

Beyond Opônê [Hafun at the Horn of Africa], with the coast trending more to the south, first come what are called the Small and Great Bluffs of Azania. ... six runs by now due southwest, then the Small and Great Beaches for another six, and beyond that, in a row, the runs of Azania: first, the so-called Sarapiôn run, then the Nikôn; after that numerous rivers and harbours, one after the other, numbers of them separated by daily stops and runs, seven in all, up to the Pyralaoi Islands [the Lamu archipelago] and what is called the Canal; from here a little more to the west, after two night and day runs, lying due west ... comes Menuthias Island [either Zanzibar or Pemba], about 300 stades from the mainland.

The *Periplus Maris Erythraei,* Anon. (ca. 100 AD). Translated by Lionel Casson.

The Zanj eat bananas, which are common there as they are in India, but the staple food of the Zanj is sorghum and a plant called *kalādī*, which is pulled from the ground like truffles, and the root of the *rāsin* ... They also eat honey and meat. ... Their islands in the sea are innumerable. The coconut palm grows in them and provides one of the foods eaten by the Zanj peoples. One of these islands called Qanbalū [either Zanzibar or Pemba], a day or two distant from the coast, has a Muslim population with inherited kingship.

From *The Meadows of Gold* by Abu al-Hasan 'Alī ibn al-Husayn ibn 'Alī al-Mas'ūdī, (896-956 AD). Translated by Paul Lunde & Caroline Stone.

We stayed one night in this island [Mombasa] and sailed on to the city of Kulwā [Kilwa Kisiwani], a large city on the seacoast, most of whose inhabitants are Zinj, jet-black in colour. They have tattoo marks on their faces, just as there are on the faces of the Limis of Janāwa [Guinea]. I was told by a merchant that the city of Sufāla [Sofala] lies at a distance of half a month's journey from the city of Kulwā ...   Its people engage in *jihad* because they are on a common mainland with the heathen Zinj people and are contiguous to them ...

The Travels of Ibn Batuta AD 1325-1354. Vol. II.   Translated by H.A.R.Gibb.

The seventh island is that of Zanzibar, extending from the coast of Zanj, covered with trees and containing rivers. In it are 40 districts which are governed by Muslim Sultans, but on the coast of the mainland above it they are infidel. ...

From the *Kitāb al-Fawā'id id fī uṣūl al-bahr wa'l-qawād'id*, of Ahmad bin Mājid al-Najdī, (ca 1480). Translated by G.R.Tibbetts.

The port of Sofala exports pure gold, and also beyond it in Kilvani [the island of Chiloane]. Do not argue I simply mention it, O man who asks me, of the coast and the gold mine - accept my indications! More to the south, O my brother, the way to these mines is hard and more than two months. You follow it. In the southern part of Sofala, after travelling for two days, without a doubt, you will find a port for all the winds, and *[the stars]* Nach *[at the height]* of 51/2 (fingers) will diminish. After this further south Nach reaches 5 fingers, as experienced people say. After that, the people of Mulbaiuni city [Moçambique], and after Mulbaiuni here start searching: later appears Malabati ahead, the one that is called "coastal land", still it does not appear. ... And later in the South appear the Charbuh islands [the Bazaruto archipelago]; there are three.

From *The Sofaliya* by Ahmad bin Mājid al-Najdī, (ca. 1500).

At sunset we anchored in front of the said city of Mombasa and we did not enter the port. When we arrived there came to us a *zavra* loaded with Moors, and in front of the city were many ships, all beflagged with their standards. We, to keep them company, did the same ...

The 'Roteiro' of the Voyage of Vasco da Gama 1497-1499. Translated by Eric Axelson.

The wind was quiet all that day and we made little progress; but the way from Mukalla to Shihr is short, and we came to the Sabaean port next morning with the baggala ahead of us – she ghosted beautifully – and the Kitami boom behind. We were twenty-four hours between these ports. At Shihr we found ourselves one of ten booms at anchorage. Eight were from Kuwaiti, and two from Persia; we were all bound for Zanzibar.

From *Sons of Sinbad* **by Alan Villiers, (1940)**

The next day we explored. It was a forlorn and strange experience because the old town was a washed away ruin in the sands of the Bay. I will always remember the particular feeling of dismay that I felt as we climbed over the brow of a sand dune to see a great low-tide plain of yellow sand before us with a small pyramid of dark stone in its centre, surrounded by black specks and pimples, like a rash. A brilliant kingfisher alighted nearby as I took photographs with my telephoto lens. We walked out on the drying sand, thankful that fate had made the tides suit us, and came to the pyramid of jumbled stone blocks that was all that was left of the Fortress of São Caetano, built in 1505. When the tide forced us to return, I searched the rubble of the destroyed town and in the sand I found part of an old square bottle of thick greenish glass, roughly moulded, with bubbles in it. It is my souvenir.

A description of Sofala from the author's book, *Two Shores of the Ocean*, **(1992)**.

I told Hassani that we would like to do the rounds of the Palace and Mosques first and we could do the fort at the end and we set off on quite a long trek, first along the cliff edge and then down to the beach, winding along a path between the mangroves and the cliff. I began to feel fatigued and wondered how long the ordeal might last, when Hassani pointed above where there were some rough grey walls. We climbed steeply up the cliff and there were the ruins of the *Husuni Kubwa*, the Great Palace of the Sultans of Kilwa.

Kilwa Kisiwani in 1998, from the author's book, *Mud, Sands and Seas*, **(2007)**.

Jahazi dhows in Zanzibar's old harbour in 1987 (Photo by the author)

Sailing in a working cargo *jahazi* off Lamu Island in 1987. (Photo by the author)

Traditionally designed *boums* under construction in Sur, Oman, in 1996.
(Author's photo)

Model of an *mtwepe*, a dhow of the medieval East African trade.

Photo : Miriam Vigar

The author at Tiwi Beach, near Mombasa in 2006

Unde etiam vulgare Graeciae dictum 'semper aliquid novi Africam adferre'.
(*Whence it is commonly said amongst the Greeks that 'Africa always offers something new'.*)
- Gaius Plinius Secundus [AD23 -79]

The sons of Africa must let the world know that we can well do without civilisation if this means that we have to throw our own culture, beliefs and way of life overboard.

- Credo Mutwa, Zulu chronicler

You cannot force the development of the soul as if it were a hothouse flower; the process must be gentle and gradual. So the true progress of Africa, in our day, did not necessarily fit in with plans for urgent economic development.

- Sir Shenton Thomas G.C.M.G., [1879-1962],
British Colonial Governor.

In this world of crowded houses, people crushed and crammed together, Hima could now believe the Dangi story, that men die only because there is no room for them all.

- Hazel Mugot, Kenyan novelist.

Glorious is this world,
the world that sustains man
like a maggot in a carcass.

- Oswald Mbuyiseni Mtshali, South African poet.

ACKNOWLEDGEMENTS

I must acknowledge the great importance of the various academics and professionals whose work I have read and quoted, and who have stimulated me with correspondence or conversation. In this book, as in its complimentary volume *Seashore Man and African Eve*, my task has been to knit together latest information and opinion with my own thinking and intuitive reasoning, all gathered together over a period of thirty years or more.

In the introductions to *Seashore Man and African Eve*, I listed those who had gone out of their way to assist me, and since the two books overlap in so many ways, I list them again.

The late Dr Richard Wilding at the Fort Jesus Museum in Mombasa; Professor Tom Huffman of the University of the Witwatersrand; Dr Tim Maggs and Gavin Whitelaw of the Natal Museum; Leonard van Schalkwyk previously of the KwaZulu-Natal Monuments Council; Professor L.B.Crossland of the University of Ghana; Prof. Jouke Wigboldus of the Wageningen Agricultural University; Prof. Felix Chami of the University of Dar-es-Salaam; Dr. Royden Yates of the South African Museum in Cape Town; Dr Tim Redfield of the Geological Survey of Norway; Prof. Phillip Tobias of the University of the Witwatersrand; Dr E Rohling of the School of Ocean and Earth Science of Southampton University; Dr Peter Jakubowski of the Naturics Foundation; Dr. Mark Maslin of University College, London; Richard Bailey of the University of Durban-Westville, and Prof. Michael Crawford of the Institute of Brain Chemistry and Human Nutrition in London.

The Killie Campbell Memorial Library of the University of Natal provided valuable assistance on numerous occasions. I enjoyed a series of courses on European archaeology at Manchester University in the early 1990s and a brief course on latest issues with Saharan rock-art at Cambridge University in 2005.

I have always found my membership of the British Institute in Eastern Africa in Nairobi, Kenya, to be rewarding. This institute organised two unique and valuable conferences which I attended. Dr John Sutton is noted for the organisation of the conference on the Growth of Farming Communities in Africa South of the Equator at Cambridge, England, in July 1994, and Dr Paul Lane that on the Maritime Heritage of the Western Indian Ocean at Stone Town, Zanzibar, in July 2006.

Miriam Vigar, besides being a staunch friend and a companion on many safaris, helped with revisions of the several texts and lent me her computer for weeks at a time when I was far from home. My companions on various safaris in Africa and around the Indian Ocean rim over some forty five years are "the very spice of life, that gives it all its flavour". I am indebted to Harry Bourne and other email correspondents who provided information and lively inspiration.

My wife Sue is always the anchor about which my restless ship swings.

NOTE ON THE SEASHORE HYPOTHESIS

In this book reference is made to two general concepts which are explained at length in my book, *Seashore Man and African Eve*.

Firstly, there is the hypothesis that mankind evolved through at least two major jumps - from forest apes to habitually bipedal hominids of the *Australopithecine* group of species, and then to the *Homo* line of descent. The *Homos* were toolmakers with the advanced physical and mental abilities enabling them to spread to the climatically practical limits of the 'old world'. It is from those first people of the *Homo erectus* species which emerged 2,000,000 years ago that we are directly descended.

It is the thesis of the **Seashore Hypothesis**, expounded in *Seashore Man and African Eve*, that it was only through the adoption of a seafood diet for long periods, with its unique nutritional attributes for the development of brains and neural systems, that these jumps in evolution could occur. Adaptation to seashore living as a natural habitat also enabled rapid migration along seashores around the Indian Ocean which greatly facilitated the peopling of Eurasia during several out-of-Africa thrusts, the last of which occurred about 80,000 years ago.

A new idea is also explored in *Seashore Man and African Eve* which is the likelihood of particular sharp evolutionary changes in culture being caused by mutations in the soft tissue which have not been detected so far in fossil bones. These mutations could have been caused by exceptional bursts of cosmic radiation from galactic sources, and including aberrations in our own sun's outputs. It is only in the last decades that sophisticated instruments, and the use made of them, that increasing knowledge of cosmic radiation makes the pursuit of this hypothesis possible.

From ice-cores it is known that there was a blazar (a concentrated jet of radiation) from the direction of Cygnus affecting Earth beginning about 100,000 years ago with later peaks at about 80,000, 60,000, 40,000 and maybe 20,000 years ago, and at those times there were particular changes in human activity; most especially the universal massive flowering of creative art and decoration by 35,000 years ago. Rather vaguer possibilities of cosmic radiation bursts, possibly from nearby supernovae detected from ocean floor sediments at roughly 4M, 2M and 500K years ago, may coincide with evolutionary jumps, enhanced by periods of sustained seafood nutrition, most especially the emergence of *Homo erectus* and the dawning of *Homo sapiens*.

It is noticeable that evolutionary jumps occurred coincident to extinctions of mammal species and diverging species or races of hominids. The extraordinary 'jump' culminating about 35,000 years ago, which is generally taken to be the beginning of the African Late Stone Age, has fascinated me for many years. Increasingly there is evidence of both a peak of cosmic radiation shortly before that time and mutation in human genes. Those two phenomena are invisible to us without the use of the latest, specialist technology and equipment but the enormous expansion - it could be called an 'explosion' - of cultural activity and range all over the world around that time has been demonstrated by archaeology for many years. Recently, my correspondent Andrew Collins passed this on to me:

> **Human brains enjoy ongoing evolution**
> 15:06 09 September 2005
> NewScientist.com news service. Journal reference: *Science* (vol 309, p 1717 and p 1720)
>
> Mason Inman :
>
> The human brain may still be evolving, new research suggests. New variants of two genes that control brain development have swept through much of the human population during the last several thousand years, biologists have found.
> The evolution of a large, complex brain has been the defining feature of the human lineage – although human brain size has not changed over the past 200,000 years. But it is not apparent whether the new genetic adaptations discovered in human brains have any effect on brain size, or intelligence.
> What is more, not everyone possesses the new gene variants, potentially inflaming an already controversial debate

about whether brains of different groups of people function differently.

"Whatever advantage these genes give, some groups have it and some don't. This has to be the worst nightmare for people who believe strongly there are no differences in brain function between groups," says anthropologist John Hawks of the University of Wisconsin in Madison, US.

Brain size

There are two new genetic studies that suggest the brain may still be evolving. Geneticist Bruce Lahn of the University of Chicago in Illinois, US, and colleagues analysed the sequences of two genes active in the brain – *Microcephalin* and *ASPM*. Both regulate brain size - people carrying a non-functioning mutant copy of these genes suffer microcephaly, where they have a normally structured brain that is much smaller than usual.

First, the researchers sequenced the *Microcephalin* gene found in 89 ethnically diverse people. The team found dozens of variants (or alleles) of the gene, but one particular set stood out. These alleles all carry a specific mutation that changes the protein the gene codes for.

This distinctive mutation is now in the brains of about 70% of humans, and half of this group carry completely identical versions of the gene. The data suggests the mutation arose recently and spread quickly through the human species due to a selection pressure, rather than accumulating random changes through neutral genetic drift.

Analysing variation in the gene suggests the new *Microcephalin* variant arose between 60,000 and 14,000 years ago, with 37,000 years ago being the team's best estimate. The new mutation is also much more common among people from Europe, the Middle East, and the Americas than those from sub-Saharan Africa.

"Compelling evidence"

The team also sequenced the *ASPM* gene from the same original sample and again, among dozens of variants, found a defining mutation that alters the protein the gene codes for. Estimates are that the new variant of *ASPM* first appeared in humans somewhere between 14,000 and 500 years ago, with the best guess that it first arose 5800 years ago. It is already present in about a quarter of people alive today, and is more common in Europe and the Middle East than the rest of the world.

"The evidence for selection is compelling," says population geneticist Rasmus Nielsen of the University of Copenhagen in Denmark. Yet it remains unclear yet how these genes work in healthy people. Many researchers doubt there is any mechanism by which nature could be selecting for greater intelligence today, because they believe culture has effectively blocked the action that natural selection might have on our brains.

Lahn and his colleagues are now testing whether the new gene variants provide any cognitive advantage. Natural selection could have favoured bigger brains, faster thinking, different personalities, or lower susceptibility to neurological diseases, Lahn says. Or the effects might be counter-intuitive. "It could be advantageous to be dumber," Lahn says. "I highly doubt it, but it's possible."

These hypotheses are referred to in the text of this book and it is hoped that a detailed explanation is not necessary for an understanding of the discussions here. The detailed explanations are indulged in my book *Seashore Man and African Eve*.

Denis Montgomery,
Chedburgh, Suffolk.
21 September 2008.

INTRODUCTION

Originally much of this book was published as part of my exploration of evolution and the pre-history of Africa in *Seashore Man & African Eve*. When preparing a new edition of that book I decided to reduce its scope to make its content more relevant and its size more manageable.

The chapters which were excised were those which dealt with African pre-history from the 'African' Late Stone Age onwards, the assumption being that by the Late Stone Age anatomical evolution had more-or-less ceased. It is the cultural evolution of mankind which has accelerated so astonishingly since then.

Within these discarded chapters were those concerned with the enormous changes which were brought to eastern Africa by Indian Ocean seatraders. There was no planned or even conscious effort; it was a by-product of cultural diffusion, by the simple influence of meeting and exchanging ideas. Trading was the engine. In a story about Portuguese maritime explorers I wrote:

> It was the results of the Indian Ocean trade, following Dias and da Gama, which were so significant and powerful. What African tribal chief cared about the political motives of strange traders from far away? Indeed, do traders ever care about such things? What mattered was their honesty, the integrity of the goods, personal manners, the conviviality of the company, the advice or amazing stories swapped around.

When writing about sub-Saharan African pre-history since 5,000 years ago one is instantly confronted with the problem of colonialism. Today, 'colonialism' is popularly equated with European activities in the last two hundred years or so and is almost universally reviled as exploitive and oppressive. Two clear facts are seldom acknowledged.

They are, firstly, that eastern and southern Africa have been colonised in three distinct phases in the last 3,000 years. The first was that of the Negroes, mostly of West-Central African origin with a 'Bantu-speaking' culture carrying their iron-age agricultural industry. They swept up the indigenous hunter-gatherers whose remnants survived only in pockets unsuitable for the colonists' cattle and cereal crops. The second was the establishment of trading city-states by Islamic Arabians and people from the Arabian Gulf who later extended their activities to plantation agriculture along the coast and on offshore islands. The third was that of the Europeans, almost entirely in the later 19th and early 20th centuries, who brought all the marvels of modern industrial technology to bear on the continent with the consequences we observe today.

The first phase resulted in the disappearance of most of the region's indigenous peoples. The second phase of a thousand years or more resulted in the Swahili language and culture, and the transportation of many thousands of native-African slaves to Egypt, Arabia and Mesopotamia. The third phase, which mostly lasted for less than a century and ended in the 1960s, is the one which is the subject of contemporary discussion and comment. It is also used as an explanation for the problems and failings of contemporary African society.

The second clear fact is that it was trade that was the driving force. The movement of Negro peoples into sub-equatorial Africa from at least 3,500 to 1,000 years ago was the result of population expansion caused by the introduction of agriculture from the Middle East, greatly enhanced by iron technology. Although it is obvious that population expansion occurred because of the introduction of agriculture, it is arguable that the spread of agriculture was related to trading activities. It is trading that spans the natural cultural or ethnic borders of population groups and spreads knowledge.

The Islamic colonial activity in Africa had several motives, but the immediate practical goal was undoubtedly trade in precious goods: ivory, metals, rare natural materials and periodic slaving. The European colonisation of Africa was focussed primarily on trade with the East, the other pole of advanced industrial production and source of rare agricultural products, principally spices. Territory was later acquired in Africa, usually by dubious treaties with illiterate or temporary tribal chieftains and sometimes by force, where sources of precious metals and agricultural produce were discovered. The British Rhodesias and the Belgian Congo are the obvious examples.

The original Negro expansion and colonisation of a huge swath of territory had no national cohesion or plan; it just happened as populations grew and the climate demanded. Neither the Arabian nor European phases had any territorial aspects until the late 19th century when the furious demands of Industrial Civilisation in the northern hemisphere, coupled to the mechanisms of that Civilisation, encouraged territorial acquisition as a requirement for the exclusive rights to exploit natural resources. Trade was the engine throbbing ever faster in the propulsion process.

Professor Tom Huffman is perhaps the master of research and scholarship of the Iron Age in southern Africa and he published his masterwork on that subject, *Handbook to the Iron Age,* in 2007. Regarding the several interlaced factors influencing the transition from the Early to the Late Iron Age in southern Africa, he wrote:

> It follows that trade wealth and political complexity also correlates closely. At the one end of the scale, Southern Nguni lacked both centralised chiefdoms and long-distant trade. At the other end, the only Level-6 capitals (Great Zimbabwe and Khami) and Level-5 capitals (e.g. Mapungubwe, Kasekete, Danangombe, Bulawayo and Dzata) occurred in the Zimbabwe trading zone. When the trade moved south to Maputo, Tsonga and Northern Nguni began to develop centralised polities.

He relates trade, and affluence beyond the needs of people created by trade, to political and social development. I have long recognised that logic and studied the proof of it.

If this book has a theme it is that the cultural evolution of sub-Saharan Africa has been driven and enhanced by external trade. It is not only the transfer of ideas and the diffusion of technical knowledge, it is the effect of these on modern mankind. Inexorably, the effect of trade is to produce a style of life which leads to social and political complexity and growth in population beyond the natural increase or decrease dictated by climate and environment. And this then too-often leads to strife, warfare, refugee swarms followed by migrations, and misery for masses of people. This process may be abundantly observed in Africa today. Formal European 'colonialism' ceased to exist in most of Africa in the 1960s and it had a noticeable territorial effect for only fifty years or so in interior Africa before then.

European territorial colonisation and the imposition of European-style administrations can no longer be blamed for the torments which can be daily observed in Africa. Native-African governments have conducted their own affairs for nearly fifty years, in many cases for longer than effective European administration operated on the ground. The torments are caused by stresses in adjusting to the rapid changes and technical advances within the greatest trading network and volume of exchange the world has seen. While this engine runs and until there is an equality of wealth and commercial competitiveness, those torments must continue. It is a dismal forecast.

Modern African populations have grown exponentially and population explosion is a symptom of this engine at work. Populations grow out of control when traditional law, ethics and the security of a people are threatened or harmed from whatever direction. It is noticeable that population growth in Africa has accelerated since the end of the European colonial period. In 1905 the population of Africa, including the Mediterranean lands, was estimated at 135,000,000. In 1955 it was estimated at 200,000,000. In 2005 it was more than 900,000,000. What more needs to be said?

Photo by the author, 2004
The mightiest monuments on Earth - at the gateway to Africa.

ONE : *NORTHERN HEMISPHERE CLASSICAL CIVILISATIONS*
Agriculture, trading and Civilisation

I have watched baboons in the wild running grass stems loaded with seeds through their fingers and eating the harvested grains. As a boy resting in the wild grassland of Africa I often did this without much thought, almost instinctively, and most of us have chewed on the stems of grasses. Domestic dogs and cats can be observed chewing grasses when some digestive awareness triggers the need. Grasses and their seeds have far-distant ancient roots as a source of essential nutrients for carnivores and omnivores.

The route to civilisation was through the stages of domesticating the most tractable of the savannah herbivores, pigs, goats and sheep, and the most nutritious cereals and fruits. The classical progression has been often enough described as starting in southern Anatolia, on the hilly plateau at the headwaters of the great Mesopotamian rivers. There is scholarly debate as to whether this is the site of the Biblical Garden of Eden, carried in tribal folk-memories and written into the Book of Genesis.

The temple site of Göbekli Tepe in Turkey, first excavated in 1994, is now revealing extraordinary proof of a complex religious centre dated to 11,500 years ago. There is a formal layout with elaborate and varied sculpted reliefs on symmetrical columns. This date is earlier than that ascribed to Çatal Hüyük (± 8,500 BP), often described as the earliest settled town related to agriculture, therefore Göbekli Tepe has been assumed to have been built by hunter-gatherers and I comment on this below. The dates are important, for they are at the end of the last Ice-age when any number of strange events were occurring.

Other early vortices of agriculture have since been defined: the Jordan valley in Palestine and the northern Nile. Wheat and barley were the pioneering cereals in Anatolia and the Jordan Valley while sorghums and millet, indigenous to Africa, were cultivated in the Nile valley and later along the Sahel.

As people learned to harvest particular grasses with nutritious seeds, inventing efficient polished stone sickles for this purpose, they discovered that where patches were repeatedly harvested, a more uniform growth with fewer weeds appeared in subsequent years. It did not need long for these intelligent Late Stone Age people to try seeding extended fields where the soil was the same and to cultivate them with hoes to promote and protect the sown seeds and kill the weeds. They discovered that provided the harvest was kept dry and safe from mice and birds the surplus could be used to seed the next crop and to provide a store against lean years.

The miraculous results of proper cultivation led quite rapidly, by positive feedback, to experiments with selection. Greater surpluses led to population explosion and cultural complexities based on land rights and extended villages and towns. A nutritional driving force promoted by the concentrated and regular diet of selected grains was at work. There were successful experiments with the refining of natural processes, such as the fermentation of beer from grains and wine from fruits. This is related to discovering how to keep milk by converting it to yoghurts and cheese. No doubt, hunters and fishermen had much earlier discovered how to preserve meat and fish by drying and smoking it. The tanning and softening of animal skins and hides for clothing would have been learned in the Middle Stone Age or earlier. Old knowledge of spinning fibres into strings and threads led naturally to the cultivation of cotton. An upward spiral of industrial, economic and social development took off, which continues to the present.

Archaeology has found a punctuation point for the beginning of the processing of grains for food. It was generally believed that the exploitation of cereals began maybe about 11,000 years ago and systematic cultivation followed. But recent research shows that the dawn of the processing of food grains occurred at least by 19,500 years ago from evidence in the Levant. Ohalo II is an important and well-researched settlement site on the Sea of Galilee which has provided details of many aspects of the life of people during the last Ice-age.

Kislev, Nadel and Carmi reported in the *Review of Paleobotany and Palynology* v.73 (1992) and wrote in their abstract:

> Charred plant remains, 19,000 years old, were uncovered at Ohalo II on the shore of the Sea of Galilee, Israel. The wild barley and other edible grasses and fruits found suggest, by their ripening seasons, that the site was occupied at least during spring and autumn. The species found provide insights into the subsistence strategy of the earliest known hunter-gatherer community of the Levantine Epipalaeolithic period. In addition, the remains of barley rachis nodes provide new evidence distinguishing between domesticated and wild types in ancient archaeobotanical material.

And from an editorial article in *Athena Review* v.4 (2),(2005):

> The archeological site of Ohalo II, situated on the southwestern shores of the Sea of Galilee in Israel, has recently provided the first direct evidence of processing of wild cereal grains . ... This includes a grinding stone with traces of ground cereals, and a paved stone feature which may represent a simple baking surface. Forty radio-carbon dates from charred plant remains have revealed an age of about 19,500 years BP. ...

The discovery of fossilised domesticated figs in the Jordan Valley, described by Kislev, Hartmann and Bar-Yosef in *Science* magazine of June 2006, dated to 11,400 years ago, suggests to them that fruits may have preceded cereals in agriculture. I do not think that there need be any competition for the first domesticated food; there was a general cultural 'jump' over the threshold to domesticating plants in the millennia previous to that time.

Current World Archaeology (No.18, 2006) reported that the figs were a 'parthenocarpic' variety which we eat today:

> Though such fig trees can appear occasionally in the wild by chance mutation, they do not produce seeds, so cannot reproduce alone. Instead, they require a shoot to be removed and replanted. ...

> ... The figs were found with wild barley, wild oats and acorns. Thus it appears that early Neolithic people mixed the domesticated figs with food cultivation, hunting and gathering.

Urban culture based on cultivation grew into fixed, territory-based civilisation. In the beginning, when numbers in cultivator communities were still small, they learned to build defendable villages and towns. Anatolia has several examples of Neolithic settlements. In that heartland of the dawn of urban civilisation, the recently-defined Pre-Pottery Neolithic period began about 11,000 BP with typical sites such as Nevali Cori, with the earliest life-size human sculpture, and Göbekli Tepe which is presently identified as the oldest temple complex in the world. Göbekli Tepe (11,500 BP) was built seemingly for the sole purpose of celebrating religious activities by the priestly elite controlling a substantial regional population.

It is believed that Göbekli Tepe was built by hunter-gatherers and this raises questions. Why should nomads, which hunter-gatherers are assumed to be, devote themselves to a fixed temple? Agriculture is not yet proven in Anatolia before 9,000 BP at Çatal Hüyük. However, the new evidence shows that cultivation was probably beginning in the Jordan valley at about 11,500BP, and it may well be that it was the rigours of climate change that provoked cultivation. I do not believe that early agriculture can be definitively stated to have started in Anatolia at the date of Çatal Hüyük . It seems more logical to me that Göbekli Tepe was built by people at the beginning of Neolithic agriculture and that firm evidence of it in that region still awaits discovery. If people were experimenting with selection leading eventually to cultivation in the Jordan valley at 19,000 BP, the practice should have penetrated to Anatolia by 11,000 BP.

Pottery first appeared in the Middle East generally at about 9,000 BP at the famous half-sunken town of Çatal Hüyük where cereals, oil seeds and nuts were exploited. At about 8,000 BP fired clay figurines were created at Nevali Cori together with sculpted stelae. Towns were built into the ground or surrounded by stone walls. Jericho in the Jordan valley of Palestine, famous for the Biblical siege by Joshua, also provides other evidence of the beginning of formal sedentary living, with earliest dates of about

11,000 BP. Jericho's first wall dates from this period. Wherever the dawn of cultivation was changing the feeding habits and nutritional regime, people discovered that they had to mark and defend territory against others who coveted their good fortune and enterprise.

It seemed obvious that during the long period of several millennia from 20,000 to 10,000 BP when mankind and the higher mammals were buffeted by severe climate change (rapid warming at 14,500 and 11,550 with ice-age cold intervening) cultural transition was occurring with advancing phases and hiatus. Simplistic categorisation of hunter-gathering nomadic or cultivating sedentary societies does not seem to be at all appropriate. It was a time of enormous and complex transition.

Before the general exploitation of fruits, grasses and other vegetable foods and fibres began around 11,000 years ago, animals had started to be domesticated. As the savannah herbivores declined in the generally cold, dry time after 20,000 years ago, the more intelligent of the hunters realised that they could gain easy mastery over some species by taming and husbanding them. They could acquire a walking larder and a daily source of milk to be drunk or processed for keeping. This was much preferable to the exhausting activity of hunting increasingly rare animals and slaughtering them into extinction. Sheep were probably the first herd animals to be domesticated, because they were smaller than cattle and more tractable than goats, and the typical fat-tailed sheep of the Middle East seem to have become ubiquitous. It was the variety being herded at the far Cape in South Africa by the Khoi millennia later. Pigs are considered to be the most intelligent of the species which man has domesticated for food. The first evidence of the domestication of pigs comes from Anatolia, at Çayönü. Pigs and various fowls are the only animals easily husbanded today within tropical rainforests.

It is not difficult to imagine how this happened. The orphaned young of most mammals can be reared by human hunters who understand their natural diet and have accumulated knowledge of their species through a thousand generations. The behaviour of different prey species was understood. When the pressures and opportunities were combined, husbanding began casually and occasionally and then, as success bred success, it became universal amongst certain people in certain areas. A herding culture evolved.

There was a precursor to this herding culture. By following protein markers, in a similar way to the genetic exploration of mankind's past and the evolution of cattle, early mutations of several breeds of modern domestic dogs have been studied. There was no single breed which was the earliest of 'man's best friend'; several varieties of archaic dogs were domesticated for different purposes in different environments. Archaeological evidence has pointed to dogs being the first animals to be domesticated and genetic studies now suggest that this occurred as early as 100,000 years ago, during the Middle Stone Age. Hunters in Africa, and later in Europe, discovered that different dogs could be tamed and used as aids and helpers to ferret out small prey and assist with the chasing and following up of larger mammals such as antelopes, horses, deer and wild cattle. Dogs are gregarious pack-animals and men learned to substitute themselves for the natural pack-leaders.

Shepherds and cattleherds did the same with flocks and herds. Men substituted themselves for herd leaders. In harsher environments, herded animals could be husbanded by moving them on to fresh pastures or by following a cyclical migration back and forth between annual seasons. Thus a nomadic, pastoral culture evolved in lands suited to that style. I see the conversion to herding from hunting to be easier than from gathering to cultivation. After all, hunters had followed their prey around as long as men had hunted, however spasmodically. Hunters had first learned how to domesticate animals, discipline and selectively breed them by rearing wild dogs as helpers and clients. Dogs were the pioneers.

In this way, two distinct agricultures evolved and became entrenched in mankind.

A third culture followed naturally as a broad spectrum of degrees of mixing of these two systems developed wherever it was practical or necessary. Probably this began haphazardly when it was widely established that trading and symbiosis was valuable and productive. Cultivators traded their products with herders and kept a few animals. Herders settled near cultivators for a time and reaped a seasonal crop. It was another masterly survival system to combat the rigours of changing or cyclical climate. In further developments of this trend, economic clientship, with all its implications leading to trading, helped people to survive after particular disasters. This third culture was a system of mixed agriculture which became the more general practice in much of Africa as later millennia rolled. Additionally, those who were masters of a mixed agricultural regime could abandon either animals or crops and survive; they could re-engage when conditions improved or they moved to suitable country.

Nomadic hunter-gatherers had learned to practice peaceful clientship in the farthest mists of time. A deprived and desperate group could seek succour from a stronger or luckier band and submit themselves for a while, or for ever, to their mastery in exchange for the umbrella of their aid. Fixed territory and urban development resulting from cultivation destroyed this ancient gentle way of survival and mutual help. Agriculturalists learned to be selfish and refuse to give assistance in the face of hardship and others learned to fight to seize food when they were in desperate need. They learned to seize 'clients' too and slavery was born.

When a balance was reached between raiding and defending, another mechanism was learned and used by those who would not join the violent way. Sophisticated long-distant trading systems were developed involving established processes understood by people who could live far away from each other. Caravans on land and ships on the sea by-passed political difficulties and old enemies learned how to regulate traders to mutual advantage. Caravanserais were provided for bona-fide traders on long-distance land routes where they and their goods could be safe for a while and markets could be held. On seashores, ports were known where seatraders could shelter and refurbish and exchange goods. The rulers who set up these arrangements learned that they could get greater benefit for themselves and their people from transit taxes and the wealth generated from trading than they could find from brigandage and piracy. Free trade has always been the best way of accumulating wealth for 10,000 years.

It may be assumed that trading, that is the bartering of goods, began well before this time. For example, at Olorgasailie in Kenya there is evidence that Early Stone Age people camped at the lakeside and established a tool-making industry in semi-permanent settlements. These people were 'craftsmen' and clearly produced far greater quantities of tools than they needed for their own use. The stone they used to make Acheulian handaxes came from rock outcrops some distance away and it may be reasonably deduced that others brought the raw rock to them to be fashioned in exchange for hunted meat or skins. Olorgasailie dates to at least 8-900,000 years ago and there is no doubt that the toolmakers were operating 500,000 years ago. There are similar examples from European sites of the same age such as Boxgrove in England where 'workshop' settlements existed. Trading began far in distant time and the practice coincided with the worldwide establishment of a creative culture inherent in *Homo erectus* which also requires a sufficiently advanced language to be able to communicate the basic abstractions necessary for the practice.

The concept of trading did not have to be learned by people of the Late Stone Age, but the extension of its practices to include more complex and sophisticated processes were learned and honed by the needs of the new agriculture with consequent urbanisation. Concepts of common values over wider areas, seasonal markets for this or that product and the establishment of market-places, demarcated areas for visitors (caravanserais), even the idea of credit with trusted trading partners from far away were established by trial and error. After literacy was invented in later millennia, accounting was an immediate use for this revolutionary skill. Some of the earliest surviving examples of literature on stone or clay tablets are accounting records. All these practices and traditions have lasted until the present because they worked.

If the concept of fixed property and urbanisation spawned selfishness and envy with the violence between groups of people which inevitably followed, trading was the peaceful relief valve which also grew from these first consequences of agriculture. Trading was learned quickly and then burgeoned to become the most dynamic engine of the newly-evolved social system which became civilisation. Trading employs the dynamic natural process of feed-back, part of the general mechanism of evolution, and is a universal law of life. Trading forces the spread of knowledge and the growth of technology and its effect is exponential.

The movement and transfer of cultural innovation leading towards a general state of civilisation was engineered by this new dynamic mechanism of trading. But it also involved some movement of peoples. The evolution of political and theocratic organisation proceeded coincidentally to the agricultural and technical revolution. All of these trends of emerging civilisation worked together and stimulated each other in forcing innovation and change. Tribal structures with hierarchies coalesced into city-states and nations with imperial dynasties.

Transfer of culture by conquest or migration is easy to understand and there is plenty of evidence from historical writing and recent centuries. The history of Britain, alone, is evidence enough. Diffusion is more difficult and its mechanism is detectable only by a combination of disciplines with archaeology leading. It would be easy to conclude that agriculture was spread westwards through Europe from the Middle East by a migrating people who carried the knowledge with them. Similarly, the spread of agriculture in Africa, westwards and southwards from the Nile and the Horn, may be assumed to have been carried by migrants. But this is not always correct. New evidence described in *Current World Archaeology*, v 2 - 6 (2006) shows that in central Europe, for example, in today's Czechoslovakia, agriculture was acquired by resident people by diffusion. Researchers from the University of Sheffield and other universities in the United Kingdom, Germany and the Czech Republic were involved in the study. The report stated:

> ... the new research indicates that it was the local hunter-gatherer communities whose ancestry can be traced back to the local late Palaeolithic, who adopted farming for themselves - through contacts, trade, and partner exchange (eg marriage), with the first farmers of south-east Europe.

Trade and contact also led to malevolent features of civilisation. The new characteristic of brigandage and theft learned at the beginning of the agricultural revolution ballooned into a concept of empire building. Tribal leaders and Kings wanted to conquer other tribes or nations to acquire their wealth or territory, acquire their technology and to enslave their people. Masses of peoples became hostage to the envy and greed of Kings or their elite establishments. Clientship mechanisms of the nomadic hunter-gatherers, living a 'natural' life, were far in the past.

* *

This general scenario works and the evidence for it has been studied by any number of scholars over many years. But, when thinking about it I have always felt that there seemed to be an awkward missing ingredient. It does not go far enough.

How did conventional evolutionary theory explain the enormous jump from reacting to climate and the regular changes of immediate environment to the forward-planning and anticipating required by all forms of agriculture? I had satisfied myself that externally induced mutation provoked the jump to creative thinking as proven by the appearance of decoration at around 100,000 BP and the explosive flowering of art from 35,000 BP.

I had necessarily to assume that the same trigger had somehow induced an ability to plan and anticipate, the necessary ingredients for agriculture and the imperative of territory and sedentary life. Somehow it happened because the evidence is there. Mutation had occurred, creativity had become a common tool of Neolithic or Late Stone Age people. But precisely how had this happened?

Clearly, as I have pointed out so often, the evidence of creative art and the use of imagination are part of a great interactive bundle of enormous behavioural or intellectual change. Not only did men begin painting on rock at 35,000 years ago and shift to sculpture at 11,000, they initiated herding and cultivation; the domestication of animals hitherto seen only as hunted prey, and plants hitherto gathered in the wild fields. Without imagination and creative lateral thinking, implying a paradigm shift in mental activity, why did Palaeolithic people, equivalent to the African Middle Stone Age, decide to begin domestication?

They had no way of visualising a great improvement to their economic affairs by domestication. They could not imagine the results of herding and a diet of milk and, more significantly, of resisting the millennia-long learned behaviour of killing prey for immediate satisfaction. Could they 'see' the advantage of organising and expending tangible communal energy in clearing land and sowing an annual crop with the expectation of a harvest months away? Annual harvesting required an understanding of seasons with an understanding of solar time in the environment of astronomical observation. Agriculture was far away from following prey animals as wet and dry seasons changed, or to remain blissfully uncaring, fishing and gathering fruits beside the tropical seashores. Not only did they have to visualise the result of herding and cultivating as economic benefits, they had to be able to anticipate and plan months in advance. Above all else, it required a disciplined society ready to accept great change and to follow far-seeing leaders. These are truly high jumps in mental activity and attitude, and in social organisation.

This was no planet-wide improvement or updating of long-existing practises like inventing better fishhooks, spears or a greater variety of stone tools. They were at the same psychic interface as the formidable one when *Homo erectus'* immediate forebears 'saw' the advantage of sculpting a hand-axe out of a lump of rock.

The mutation which occurred at 35,000 BP began the process which came to fruition with the next burst at 11,000 BP. Artistic flowering and the emergence of tiered religious intellectual vision are symptoms of the effects of the mutations in the brain. They were genetic imperatives.

The work of Professor David Lewis-Williams extends quite naturally into the coincident extension of massive culture change involving the emergence of agriculture and the attendant attributes of territory and civilisation.

He proposes, along with the rest of his thesis, that it is necessary to theorise an intellectual fabric complementary to that which gave birth to religion and structured society. Proto-Neolithic people in the Middle East did not divide the environment into 'wild' and 'domestic', because they had not yet made that distinction. All animals and plants were part of the same cosmos as people. There was no divide between hunter-gathering and agriculture, it is we who have learned to divide the world between wilderness and civilisation, or the wild and the domestic.

The cosmos was tiered and the supernatural or psychic aspects merged from the alert or Earthly state, through a psychic vortex into cognisance of an underworld and a heaven, in deep dreams or trance. A view of the physical world was also tiered; mankind was at the peak, animals followed in order of intelligence to the plant kingdom. The Book of Genesis spells it out clearly enough. The paradigm shift in the 'hot-wiring' of the brain, using Lewis-Williams' analogy, enabled people to imagine and plan. Their understanding of a completely 'whole' cosmos of all life enabled them to exploit the domestication of animals and plants without difficulty. They did not have to try to cross an intellectual bridge between hunting and conserving, gathering and sowing. There was no great intellectual exercise or change in cultural practice to decide to nurture orphaned antelopes or sow the saved seeds of preferred cereals. Once the 'hot-wiring' had occurred resulting in the ability to foresee and plan, the practice followed naturally.

To explain the motives for this extraordinary change, Lewis-Williams proposes that a factor in the domestication of herd animals, if not the principal factor, is that men began to keep them for social and religious reasons. Keeping domestic animals showed that the owner had power over lower animals in the tiered cosmos. A herd proved prestige and status. There was a religious function in the use of domestic animals in sacrifices to gods and ancestors. Animals were part of increasingly complex marriage customs and symbolic tribute from subservient persons to overlords. They would only be eaten during ritual celebrations; but they did provide abundant year-round nourishment in the form of milk and processed yoghurts and cheese.

David Lewis-Williams and David Pearce write in *Inside the Neolithic Mind* (2005):

> Domestication [of wild cattle] no doubt did lead, later if not immediately, to easier availability of milk, animal fibre and so forth. However, whether greater security of production was also attained is a moot point; even though they can support higher populations, domesticated animals are more susceptible to disease and the vagaries of nature than are wild animals. [I would also add that they become property and therefore become a source of envy with vulnerability to theft and, therefore, violent conflict.] A desire for secure production was not necessarily the reason why people tried to domesticate animals, as many accounts imply.

What is noteworthy is that these are precisely the ingredients of the 'cattle cult' which remained deeply entrenched in the culture of all African herders and mixed agriculturalists until the present day. The detailed history of native-Africans and their relationships within rival tribal societies and with European settlers is dominated by the importance of cattle as symbols of wealth and status. Many historically recorded wars and skirmishes occurred over cattle herds in southern Africa.

In discussing the reasons for the shift from gathering to cultivation, other complex but similar factors have to be considered.

Cultivating requires an understanding of seasons and the knowledge to be able to forecast in time. Together with the ability to visualise tools within lumps of rock, an understanding of time, I believe, began in the *Homo erectus* phase of human evolution because of the necessary daily study of tides during many millennia of seashore living. (See the Author's Note, and it is discussed in my book, *Seashore Man & African Eve*.)

Diurnal tidal variation came to be understood since it was important to daily existence. This knowledge led to an understanding of the connection between spring and neap tides and the phases of the moon. The daily passage of the sun and the regulation of the ocean by the moon progressed to an awareness of the ponderous movements of the starry sky by early *Homo sapiens*. However dim these understandings were, the mutated 'hot-wiring' of brains during the period of varying intense cosmic radiation from about 110,000 to 11,000 years ago progressively sharpened their intellectual ability. (See the Author's Note, and this is also discussed at length in *Seashore Man & African Eve*.)

Intellectual or psychic understanding of the inner cosmos of the mind was integrated with an extended bundle which included awareness and study of the 'real' cosmos of sun, moon and the slowly wheeling stars. Time became important, not just in regulating gathering on ocean reefs, but on an annual scale. An understanding of annual cycles, demonstrated by the solar seasons and reinforced by the circling constellations, were all part of that big creative jump at 35,000 BP and soundly reinforced 11,000 years ago. Astronomy, integrated with astrology within the cosmological whole, became a vital intellectual tool in the hands of priests and rulers. This is particularly true of those communities and city states whose cultivation cycles and thus survival depended on the annual flooding of rivers providing irrigation, as desertification proceeded in the Holocene. Mesopotamia, the Jordan and Nile valleys are where cultivation and resulting urban civilisation matured. The Indus and the great river valleys of China were not far behind.

David Lewis-Williams and David Pearce:

> In place of ecological imperatives and ineluctable forces of capitalist optimization, we point to the building up of social status and so link the domestication of animals to aspects of human history, cosmology and myth. There

was, we argue, a creative, dynamic interplay between the cosmology and imagery of the Neolithic, their social concomitants and domestication of animals.

This transition of the way in which animals were viewed and then managed, born of the mutational 'hot-wiring' of the brain and mind, which becomes consolidated at 11,000 years ago, also produces the concept of property. Herds became property of clans or individuals whose status depended on their size. The same principles applied to cultivation. A store of grains was more than a larder, it was part of the "interplay between cosmology and imagery" obtained by power over the plants. Cultivating, especially with irrigation, requires considerably more energy, both physical and psychic, than gathering wild plants and fruits. The construction of fixed towns where land was cultivated, and the ordering and defence of them, required enormously greater expenditure of energies. The imperatives born of the cultural jumps experienced exponentially after 35,000 years ago clearly had other internal sources; they were not the result of environmental pressures. Those pressures had been accommodated by the *Homo* line of descent for hundreds of millennia.

It was a genetic imperative, 'hot-wiring' by mutation born of cosmic radiation, that transformed society.

*   *

The severe desertification of the Sahara 25-12,000 years ago had propelled people to the periphery and the Nile valley was a magnet. However, they were not the founders of the fixed Nile valley communities. Sedentary Late Stone Age fishermen and hunter-gatherers resided in Egypt for many millennia long before the evolutionary jump to a sustained agricultural economy and the intense revolutions of Civilisation. Migrants from the drying Sahara joined the settled communities. After the agricultural revolution, stimulus from the Middle East was continuous with communication and migration back and forth via the Suez land bridge.

Suddenly, a complex web of invention and experiment, based on surplus of food and the leisure that follows from wealth amongst the elite classes, seemed to explode in those areas where cultivation was most successful.

The megalithic monuments and buildings of the first civilisations have always fascinated modern travellers and scholars. Many aspects of their construction still puzzle experts at the end of our 20th century. The design and construction of pyramids and temple complexes all over the northern sub-tropical zones of Earth, are awe-inspiring to us despite our ability to build rockets which reach the outer planets and steel, glass and concrete towers that climb into the skies in every modern city.

What is perhaps the most fascinating and significant fact related to the early civilisations' architecture and civil engineering industry, is the perfection of their mathematics used in design and the accuracy of their execution. The precision of the layout and construction of the great pyramids of Giza in Egypt commands awe in any thoughtful modern architect or construction engineer. Their orientation to astronomical signs in the heavens, their use of sophisticated mathematical symbolism and the extraordinary feats of megalithic construction prove that their designers and builders had an order of abilities not much bettered in 5,000 years.

Lifting numerous 200 ton immaculately quarried and carved single blocks from virgin rock, cut to tolerances of a few millimetres and less over several metres in length, is a challenge for any engineer without computer guided mechanical tools today. Transporting such enormous blocks over miles of rough terrain, erecting them and joining them without the aid of mortices or cements with a stability to last thousands of years and withstand many earthquakes is daunting.

The particular genius and achievements of the megalithic designers and constructors of the great ancient civilisations were not carried into the interior of Africa away from the vicinity of the ribbon of the Nile. They were neither relevant, necessary nor possible. Life away from the Nile was hard and it is still hard today.

The simpler manifestations of agriculture, coupled to intellectual flowering, were sufficient to create a huge jump in cultural evolution for the mass of Late Stone Age people living outside the Nile valley.

The Sahara and southwards

In the Nile valley there was population pressure and movement outwards across the Sahara of savannah and up the great river road. There were also the ancient pathways along the seashores which hominids had used for a million years.

Roland Oliver wrote (1991) :

> By about 9,000 years ago the Natufians of the Jericho district [Jordan valley] were living in brick-built settlements that may have housed as many as 2,000 people, and were harvesting cereals that showed signs of deliberate modification towards the domestic forms. Egypt lay on the southern periphery of this area and it was only around 7,000 years ago that it received the benefits which by this time had been extended to include various legumes, such as peas, beans and lentils, and also domesticated species of goats and pigs, sheep and cattle.

This does not imply that the Jordan Valley was the sole innovator of cultivation in the area; it is an illustration of the interchange of ideas and products. There was proliferation of agricultural know-how and the results of selection and experimentation were shared by trade and its mechanisms of communication. Whereas wheats and rye grasses were developed in the cooler regions such as Anatolia, sorghums and millets became the staple of the Nile. These cereals were better suited to tropical climates and their parasites and were exploited by cultivating migrants as they moved southwards.

When continuous cultivation of cereals and new breeds of domesticated animals had become established in lower Egypt, some Saharan people were already versed in systematic harvesting of grains and managing indigenous sheep, ancestors of the mouflon, ready to receive the newly selected and cultivated breeds from the Middle-East. They were also ready to adapt and modify fixed village cultures of square mudbrick houses, pottery industry for storage and cooking and the disciplined constraints of complex communal living. Following selective gathering of cereals which had been practised probably from 11,000 BP, at about 7,200 years ago the herding of domesticated animals is detected by archaeology. This was the beginning of the Pastoral Period of the Sahara with changes to rock-art styles and the definition of camp sites. Pottery relics and stone tools are found abundantly.

The simpler manifestations of agriculture, coupled to intellectual flowering, were sufficient to create a huge jump in cultural evolution for the mass of Late Stone Age people living outside the Nile valley.

About 3,500 years ago cultivation in the Sahara had become quite widespread towards the end of the wet period. Grains and cereals need good storage and have to be cooked in pottery vessels, and this was another stimulus to inventiveness, industry, social cohesion and the idea of trading. By 2,700 BP, the Garamantian culture had begun with evidence of permanent settlement at oases, the cultivation of date palms, irrigation, widespread use of horses and the wheel. As desertification progressed horses were replaced by camels. All of the domestic animals, whether for food and wealth or for transport, were imported already-domesticated species from the Middle East, either from Egyptian sources or from Arabia via the horn of Africa. Rock-art shows that woven cloth was used as clothing and there are the ominous signs of warfare with the depiction of armed figures. The benefits and curses of civilisation had arrived.

The use of coarse clays mixed with sand for house-building was another most significant jump in technology. The particularity of square, brick-built houses leading to rectangular town layouts is a clear signpost that the communities building in that way had direct cultural contact with the agricultural Nile Civilisation. The significance of square houses in a grid of streets or alleys is clear. Square buildings and layouts make it easy to squeeze buildings into a small area. Space was not a problem, but defending a community from aggressors was. The more concentrated the urban area, the larger the defending force could be relative to the population. Square houses and towns with a grid of streets is a product of a sedentary lifestyle with all the problems of a territorial and materialistic civilisation.

Pottery for the storage and cooking of cereals were essential tools of the farmer and town-dweller. Without a pottery industry, cultivating has no purpose; the two have to come together. Since the availability of clay is not universal, potters became experts and pots were an early product suited to trading; especially pots which also performed a dual purpose as packaging for other liquid or loose trade goods. Firing good clay resulted in waterproofing and tough pots which could be used to store liquids, and for the essential cooking of cereals and tough roots. Having regard to the inherited creative drives of the Late Stone Age, it is not a surprise to know that the decoration of pots for aesthetic reasons, like the early eastern Saharan wavy-line design, was followed with distinctive designs for identification of a communal culture.

In time, as the concept of property grew with the needs of establishing ownership of agricultural lands however temporarily, pottery designs became formalised and the mark of communities and then whole societies. This became so entrenched that archaeology today relies heavily on pottery designs to trace the ancestry, movements and affinities of tribal groups and broader linguistic or economic communities. Pottery, a child of agriculture, became an essential part of all cultivators' lives, not restricted to urban society and the rising civilisations along the Nile. Pots can be made by anybody with simple learned skill and practice as long as there is clay and water available. Pottery moved everywhere with farming.

Professor Tom Huffman in *Handbook to the Iron Age* (2007) makes these general remarks:

> Because pottery was an active part of culture and a representative part of the larger style [of communal art], it can be used to recognise groups of people in the archaeological record. Two conditions must be met to increase the validity of its provenance, The ceramic style must be complex otherwise it will not be uniquely representational. And secondly, the makers and users must belong to the same material-culture group. ... [But, material-culture groups] usually do not represent entities defined by blood, such as lineages, totems or clans. ... Material-culture groups do not necessarily represent political organisations such as tribes or chiefdoms.

In tracing the spread of agriculture, pottery may be a key. Pottery was made in the Sahara nine thousand years ago and this is consistent with an understanding that pastoralism was practised there by that time. Similar pottery was made in the Turkana area of northern Kenya then too and it may be assumed that Late Stone Age agriculture had been adopted by some people south of the Sahara.

During the height of the wet period following the end of the last ice-age, a fishing economy using bone hooks and harpoons had been common from Turkana to the upper Nile and all along the well-watered Sahel with its vast lakes and floodplains, and along the Niger River. The dynamics of climate change at the end of the ice-age undoubtedly increased populations and agriculture and its associated culture may have moved with migrations. It may also have moved ahead by diffusion; people learning from each other and from traders seeking new sources. It is clear from archaeology that there was movement of culture in both directions.

But where permanent and fixed towns were established, together with all these developments in agriculture and the necessary adjuncts of industrial technology for the provision of tools, weapons, household utilities and house constructions, there was a burgeoning of intellectual development. Mathematics, astronomy, writing, accountancy, legal systems, organised religion and the arts: all had to keep pace with the natural population growth in urban structures. Otherwise there would have been social and political chaos.

The migration of people armed with agricultural and intellectual refinements spread outwards across the Sahara and into the Horn during the Holocene Wet Phase and mixed with those already living there. Not only did this occur in an east-west axis, agriculture spread southwards too, into the Great Rift Valley and its many lakes and the plains of East Africa where climate change had created a congenial environment.

Where there was no migration and physical mixing, further and further from the Nile, culture and technique was transferred by diffusion in the way that it has always been since *Homo sapiens* emerged. Most usually, I believe, it occurred in a 'mosaic' or matrix of both methods. Scattered bands met, news and ideas were exchanged, young people seeking mates found them in other groups, examples were copied.

While megalithic civilisation with its several dynastic political structures and technical eras flourished along the Nile, African savannah people consolidated the three principal agricultural economies.

There were people who adopted a completely sedentary life, based on locations which supported fishing and cultivation year in and year out. Irrigation techniques were designed to ensure the stability of these farming areas. Where there was insufficient water for cultivation, there were nomadic pastoral herders who moved with their cattle or sheep as the rains dictated, relying on gathering for additional subsistence and barter for artifacts. The logic of using the dairy products of their herds and flocks rather than slaughtering them for everyday food developed naturally. Milk and processed milk products became the staple produced by their animals, meat was a luxury attendant to ritual and celebration. And there were people who developed mixed agriculture in median climate zones where they cultivated gardens and kept some domestic pigs, birds, cattle, goats or sheep, and relied to a greater or lesser extent, from season to season, on fishing and hunter-gathering.

Blurring between these different economies and cultures has occurred throughout Africa ever since, for the last 5,000 years at least. The modern Tswana of Botswana and South Africa are good 19th century examples. Professor I. Schapera of the University of Cape Town in his classical study, *Married Life in an African Tribe* (1940), described life before and after the intrusion of modern Western Civilisation in the Tswana-speaking Kgatla tribe. Though the book is primarily an anthropological study of social relationships, it is an extensive portrait of a typical mixed-agriculture savannah community at the culmination of the indigenous African Iron Age. The lifestyle of the Kgatla was probably not very different from that of their ancestors at the dawn of the Iron Age in Africa 3,500 years ago.

The Tswana established large villages or towns in suitable locales with cultivated lands around them, ruled by dynastic leaders from aristocratic families. They also had large herds of cattle and small stock which were moved about in the range surrounding the settlements. The women tended the fields and the men looked after the cattle. Hunter-gatherer San-Bushmen lived in symbiosis or clientship with these modern Tswana, trading hunted wild meat, essential herbs and their labour as herders for metal artifacts and sustenance during prolonged droughts.

By three thousand years ago, these new systems, increasingly backed by metallurgy, had worked their way down to the Sahel region of West Africa, the southern Sudan and Ethiopia. This was the Garamantian period in the Sahara when people were sedentary, living along rivers and in oases, cultivating grains and date palms, husbanding sheep and cattle. They had horses and chariots. The characteristic square house of the Middle-East and Egypt made its appearance around Lake Chad and progressed westwards. Aesthetic pottery designs developed until they could distinguish culture and tribe. Agricultural technique was applied to indigenous plants and where Middle Eastern wheats and barleys could not survive, native millets and sorghums sufficed. On the fringes of West African and Congo rainforests, people increasingly turned to cultivation of root-crops and the regular harvesting of oil palms.

The drying of the Sahara weakened the strength of the Garamantian people and the spread of civilisation southwards had to run out of steam in the tropics. Exotic horses and cattle breeds from the healthy seasonal northern grasslands could not survive diseases carried by flies, ticks and other parasites in the tsetse-infested bush and forests of tropical Africa. Plants that thrived in irrigated fields beside the Nile died from the ravages of armies of tropical insects and the insidious spread of fungi. Africans learned 3,000 years ago that it is not easy and often disastrous to introduce new crops and exotic domestic animals into the tropics, and Europeans had to re-learn this during the 19th and 20th centuries.

*

Copper is the bulk component of bronze and brass and apart from deposits in Egypt this metal was scarce north of the Equator in Africa, so there was never an African Bronze-age outside the Nile civilisation. Vast resources of copper in the far west in Mauritania, and in southern Africa in the Congo, Zambia, Zimbabwe and South Africa were not exploited until after the Iron Age commenced. Until the knowledge for iron smelting, which is more difficult because of the necessary higher temperatures and greater technical care, was acquired, sub-Saharan people had to continue to rely on bone, wood and stone tools. Trade in metals which sparked further explosions of civilisation in the Mediterranean, the Middle East and India came late to Africa, but its power could not be stopped.

The Sahara may have filtered the spread of urban civilisation and earliest examples of metallurgy but the Nile provided a highway to the Sudan and Ethiopia. From the fertile mountain region in southern Arabia other influences penetrated the Sudan and Ethiopia as regular navigation of the Red Sea commenced 5,000 years ago. With the beginning of modern trading systems, social and cultural traits and ideas travelled together with agricultural and other techniques. By 3,500 years ago, Red Sea navigation and the inclusion of the Horn of Africa in regional trading was established. Ideas and goods began entering East Africa by several routes: a general diffusion across the Sahara itself to the Sahel lands, directly up the Nile, from the Red Sea where new states were growing along the shores and the watered escarpments of south-west Arabia to Ethiopia, and by sea itself as navigation commenced between the Red Sea, East Africa and Mesopotamia and thence to India.

The concepts of wealth and property combined and interrelated with the need for order and some form of law resulted in tribal systems, ruling dynasties and a growing sense of history and dynastic ancestry. Awe of the night sky and study of it generated an awareness of a limitless Universe and some unimaginable power, or God, that drove an age-long regulation of the seasons. Wonder about origins of this Universe, life and death interacting with the aesthetic imaginations of Late Stone Age minds were codified into religions supervised by a parallel priestly hierarchy usually combined with the chieftaincy.

Systems of delegating power through clan-heads developed and with them, driven by the problems of administering wealth, rules of inheritance and family structures became entrenched in tribal lore. All the complexities of a modern civilisation, however poor or primitive the economy might seem, had to be reasoned through, adopted and accepted by the various peoples engaging in the different economies in their different environments.

When increasingly rigid tribal organisations formed progressively across Africa, Africans were following the general order and system of civilisation which began emerging first in the Middle East. The real or apparent material and technical wealth or poverty of a people at any one time cannot determine their level of civilisation. In my view, the only people in Africa after 2,000 years ago who were not 'civilised' were the hunter-gathering Pygmies and Khoisan surviving in pockets beyond the reach of those eternally tethered to their herds of exotic domestic animals or caught by the tyranny of the hoe and sickle.

*

The discovery of iron smelting along the Sahel and the Interlacustrine Zone of Central Africa in the first millennium BC exacerbated all of the attributes of Neolithic agriculture and sedentary living. Iron tools enabled the rapid clearing of forest lands and the exponential increase of both cultivation and human populations. It also enabled the keeping of exotic domestic animals in cleared areas by the coincidental elimination of flies carrying life-endangering parasites. Another lesson was learned. An upward spiral began.

Severe territorial conflict leading to warfare began south of the Sahara where I believe it had never occurred before. Leadership contests, personal disputes and family squabbles in a gregarious society often progressed beyond argument and to displays of violence, but it was occasional and shocking, causing social trauma and unhappiness. Hunter-gatherer bands disputed territory, especially in bad times, but millions of years of evolution had produced the necessary social mechanisms for resolution without resorting to self-destructive genocide.

Clientship was the method that long time had produced as the only proven method of survival when life was considered more valuable than pride. Indeed, it may be seen that the folly of pride, which has caused death and genocide as prolifically as envy, is a product of property-ownership and was not of much importance to hunter-gatherers. Does a chimpanzee or baboon suffer pangs of psychic pain when daily outfaced by a superior male or female in the hierarchy of the home band? I have watched a brash young male baboon repeatedly challenge an alpha male and be vanquished with equanimity. Why should primitive mankind have been any different until there was property to defend or covet? Robert Ardrey cited the *Territorial Imperative* as the source of human violence and warfare. In that he was perfectly correct, his error was to attribute it to primitive hominids and not to the advent of fixed agriculture and urban civilisation.

The wind of change from northern civilisations brought warfare, an entirely different order of violence. In war, institutional violence is organised by the clan or tribal authority and complete strangers are set to kill and maim each other. The particular brutality that accompanies war is caused by this unnatural behaviour, conflicting directly with our genetic inheritance. In order to wage war, extraordinary measures had to be taken to overcome the instinctive aversion to violence that all people had developed during the previous millions of years of evolution.

Chimpanzees and gorillas have been quoted as waging planned aggression, which suggests that it is endemic in all higher primates. Recently, in West Africa, chimps have been watched and filmed in organised hunting forays against monkeys for food. What is not often considered is that our ape cousins are now going through an almost insupportable trauma in increasingly restricted habitats in reserves and are being trapped and slaughtered into extinction by human hunters. There is nothing 'normal' or 'natural' about the environment in which those chimps and gorillas are being studied.

When Jane Goodall first began studying chimps in the Gombe reserve on the shore of Lake Tanganyika, they showed aggression towards her when she approached too close and there is one occasion which she relates in detail (*In the Shadow of Man*, 1971) when she was surrounded by a group. But they backed down and left her alone. Dian Fossey had similar experiences with gorillas before they got to know her but was never harmed because she always showed submission to the powerful alpha males. Goodall described encounters between chimps and baboons when both sides were combative, but organised group aggression was lacking. It is in more recent years that properly observed pre-planned group belligerence leading to bloodshed or death has been recorded.

Perhaps researchers have been observing cultural changes in our nearest ape cousins which we endured 5,000 years ago and which were caused by similar pressures: increasingly restricted territory and disruption to traditional subsistence and social structures. During the Angolan War in the 1990s, the South African Army successfully recruited and trained a regiment of peaceful San-Bushmen for active combat service proving that aggression can be quickly taught within a socially disrupted and confused group.

The adoption of sedentary cultivation leading to civilisation created territorial pressures on people, as if they were creating self-imposed game reserves around themselves. Instead of hunter-gatherer bands sitting around a fire to argue away a problem, conscious that all were pitted against the same enormity of nature and geography, part of the same Universe, people had now to defend their tribe's real property in a world of expanding populations of both people and domestic animals. When the Sahara began desertifying, 2,500 years ago, acute and insoluble problems of territory and property arose.

Organised religion had to develop further alongside chiefly power with spiritual and physical penalties for rebels and infractions of law to maintain order. Particularly, it was necessary to motivate and sustain conscripted armies engaged in impersonal battles and controlled genocide against people with whom they had no personal conflict or even contact.

War dances evolved and religious activities were invented to stir the martial spirit and instil the necessary disciplines for the cult of battle. The trance-dances which Late Stone Age hunter-gathers used for healing and the creation of harmony in a small group were extended to marshal the bloodlust in a thousand chanting warriors and their supporting ululating women.

With the burgeoning of civilisation in the wealthiest and most advanced nation states in Mesopotamia and along the Nile and the commencement of dynastic and territorial warfare, the social scourge of slavery which developed from age-old mutual-help clientship was harnessed to provide cannon-fodder. Captives and the families of captives became property and servants under laws not much different to those governing cattle. Slaves were disciplined into expendable armies. This practice continued in Africa into the 19th century where it was notoriously used by Nguni generals and clan chieftains from KwaZulu-Natal and described in the historical record.

Born in Egyptian civilisation, another social mechanism evolved to formality in this period; the concept of a totem affiliation which transcends clan or tribal boundaries and which can be used to link 'lost', enslaved or nomadic people. At initiation ceremonies, groups adopt a totem, usually an animal with whom the group or their teachers sense communion. It is often a bird. That totem then becomes taboo for that group and those with that totem have a religious or 'spiritual' kinship with all others with the same totem. Africans meeting maybe thousands of miles away from their homeland would ask about a stranger's totem and if it was the same, there would be instant kinship. In any case, there would be recognition and understanding of each other's status and background. In Western Civilisation this same practice is common, if not in such rigid and universal formality. We talk of the 'old boy' and other 'networks', people ask each other what Zodiacal sign they were born under. The roots of these comforting practises lie deep in Africa and may have been born long before civilisation spread when hunter-gatherers roamed freely in a relatively empty continent.

The totem system which made the totem animal taboo, aided a natural conservation of the environment. As Lyall Watson in *Lightning Bird* (1982) has pointed out, men of a totem could not kill 'their' animal and this helped to preserve that species. Elsewhere, I have emphasised the fact that Africans have always been selective and careful hunters until the advent of guns and the degrading of hunting practice by immigrant Asians and Europeans. In 4,000,000 years, Africans did not consciously cause the mass extinction of any prey animals.

If Africa was an Eden 35,000 years ago when art and abstract thought began to flourish with the Late Stone Age in balanced non-violent, scattered society, it lasted only until 5,000 years ago. Civilisation brought enormous change and all kinds of intellectual flowering and benign cultural evolution, but it also brought the stresses of urban living, population explosion, easily-transmitted epidemic disease, degradation of the environment and mass slaughter of wild animals, periodic mass famine and the horrors of organised warfare and slavery; all of which continually provoke mankind to expand populations beyond sustainable levels.

In *National Geographic* of August 1993, Dr. S. Boyd Eaton of Emory University is quoted as stating: "The life-style for which our genetic makeup was selected was actually that of Stone Age foragers." Eaton had been conducting research into the 'surprisingly' low incidence of cancer amongst San women of the !Kung group in southern Africa. His statement is true and obvious enough; what is interesting is that scientists and serious journals still found it newsworthy enough to announce it.

I believe that Adam and Eve's apple was undoubtedly the mutated fruits of the Middle East which responded so well to cultivation and hybridisation. Who was the serpent? Perhaps it was a blazar of cosmic radiation from the direction of Cygnus of about 40,000 years ago?

TWO : *A VORTEX IN EAST AFRICA*
Eastern central Africa, an entrepôt for cultures.

A few miles outside Nakuru in Kenya, overlooking the lake famous for its flamingoes and the game reserve in the surrounding wilderness scrub and patches of forest, there is a rocky knoll called Hyrax Hill. Supported by financial contributions from local British settlers, Mary Leakey excavated it in 1937 and in later years. It has been excavated several times since by others. It is important as a site where both Late Stone Age and Iron Age farmers established homesteads.

Homo erectus of the Early Stone Age occupied this part of the Great Rift Valley and there are sites nearby where deposits of Acheulian hand-axes have been found. Kariandusi near Lake Elementeita, dating from a million years ago, is one of these which tourists may visit. The transitions from the Early Stone Age to the African Middle and Late Stone Ages have been charted through archaeology in the area.

During the wet cycles of the last 10,000 years Lakes Nakuru and Elementeita expanded massively, forming one large sheet of fresh water and Hyrax Hill became an island or promontory and was probably submerged from time to time. Although no artifacts or remains from that far back have been found, the excavations at Hyrax Hill have provided detailed 'snapshots' of people living there at two important points of time in the last three millennia. Further archaeological exploration may provide more snapshots because there are several burial mounds and hut circles.

Hyrax Hill was a pleasant place on which to live over many thousands of years, depending on the state of Lake Nakuru and the climate in general. An old European settler farmstead, rough and simple with a verandah and corrugated-iron roof, shaded by exotic jacaranda trees and brightened by red poinsettia, served as the site museum.

The importance of the earliest Hyrax Hill snapshot is that it shows that about 3,500 years ago, people cultivated cereals and buried their dead with care, the females accompanied by their personal stone pestles and grinding stones useful for the after-life. The graves were well-fashioned with stone lining and covered with flat cap-stones. Associated with these Late Stone Age burials there were occupation sites with various artifacts: obsidian tools, waste flakes and stone cores, more platters and grinding-stones, pottery and bits of domestic animal bones and teeth. Those people were mixed farmers with a pottery industry and a culture which included ritual burial. However primitive their economy and far from the sophisticated Nilotic society of that time, the general culture of northern Africa was pervasive into the heart of what is modern Kenya.

In the more recent snapshot, between 400 and 200 years ago, people sunk stone-lined byres for calves and small stock. Huts were alongside these byres with stone foundations and gate posts with a wooden superstructure, whether clad in clay or grass thatching, and a thatched roof. There were skeletons in graves, iron tools, pottery, jewellery, domestic and wild animal remains and smoking pipes for marijuana or other herbs. Imported glass beads and Indian coins dated from 500 years ago proved occasional trade with the distant Indian Ocean.

Bau gaming boards were cut out of the living rock on Hyrax Hill. They come from the Late Iron Age period. The *bau* game, best described as a rapidly executed and complex blend of checkers and backgammon requiring considerable skill with mental arithmetic, has local names but common basic rules and is played throughout sub-Saharan Africa and across the Indian Ocean. *Bau* is the name given wherever Swahili penetrated from Somalia to Malawi. In West Africa the name is usually a variation of *warri* or *oware* and a similar game has the same name as far away as Malaysia and Singapore. On the Comores Islands between Mozambique and Madagascar it is called *mraha*. Ugandans call it *mweso*. In Zimbabwe it is called *isafuba*. In nearby South Africa, people from three different language groups have a similar name for the game; the Swazi call it *intjuba*, the Venda *ndzichuva* and the Tsonga *chuva*.

An archaeologist friend discovered a sculpted board out of soapstone in an Early Iron Age dig in the Thukela valley of South Africa. I have watched it being played from KwaZulu-Natal to the Niger River. I have seen a similar game with a pattern of concentric squares rather than hollowed hemispheres played by soldiers in a fort in the Indus Valley in Sindh and the same variation played in the sand at Inhambane in Mozambique.

Board games may be pivotal to an understanding of the emergence of common cultural activity across large portions of continents where there is no apparent cultural relationship. Irving Finkel of the British Museum provided a broader picture of ancient games in an article in *The Illustrated London News* in 1990.

There is always doubt and controversy about the connection between similar culture without apparent direct ancestry or obvious physical contact between peoples. There are a number of examples: the emergence of particular artifacts such as musical instruments and weapons, the apparent coincidental invention of pottery or an artform, common irrigation and other agricultural techniques, similar dress or adornment in widely disparate communities and so on. In African anthropology, there is dispute between the concepts of transfer of knowledge or culture by the migration or absorption of peoples and the diffusion of ideas through propinquity or trade. I have always assumed that both methods are common and very often coincidental : trying to tie down human activity to a black or white scenario is a useless and purposeless task.

A study of board games helps to illustrate that quandary. H.J.R.Murray in *History of Boardgames other than Chess* (1952) identifies five categories of which the Afro-Asiatic group is one which he names *Mankala*. This group includes the various games I have listed above such as *warri* or *bau*. Murray's other four categories are 'alignment games', 'hunt games', 'race games' and 'war games'.

Amongst 'hunt games', modern Solitaire is an example of ancient derivation. In the category of 'race games' there is the 'Game of Twenty Squares' which was played about 3300BC in Babylon, has been identified throughout the Near East and which was still being played in Cochin in western India by an isolated Jewish community in 1900AD. *Senet* is a game known by Egyptologists from Pharaonic times and identified with *Tab-es-Siga* still played along the Nile as far south as Sudan 3,000 years later. Other common examples of games with ancient roots are Ludo (derived from the Indian game, *Pachisi*), backgammon and 'war games' such as chess.

It would seem that board games have been passed on from generation to generation for millennia and jump across lands and continents for all the reasons that modern people have communicated in the last 30,000 years; through migration, nomadism, conquest and clientship, forced movement of enslaved communities, and possibly the most important of all: cultural transfer by traders or tourists. An article in the journal, *West Africa*, in September 1996 gave a useful overview of the *warri* game, speculates on its origins and its worldwide spread in modern times. Various local names for the game are quoted from Antigua to the Philippines and it is proposed that it is the world's oldest game and that it originated in Africa. I have no argument with that. It is probable that the *warri* game spread outwards from Africa, carried by the advent of trading and increased intercontinental travelling which followed the end of the last Ice-age and the beginning of civilisation.

The Great Rift Valley surrounding Hyrax Hill near Nakuru narrows and is borne up by ancient deep seismic pressure. It is spectacular scenery of steep escarpments and lurking volcanoes with equable temperatures all year round and enough rainfall to promote prosperous agriculture. To the south, where Olorgasailie lies, and north beyond Nakuru, the floor of the Rift both widens and drops in altitude and the combination restricts the rainfall. Only scrub bush survives without irrigation and natural cultivation is only possible on some flanks and on the heights behind the escarpments where rain is culled from the monsoon clouds in their seasons.

The East African monsoon provides a bonanza: where rain does fall, particularly over mountains and highlands, it rains twice a year. There are two rainy seasons which is yet another reason why that part of Africa was always blessed for mankind and all life. No wonder the Masai, Kikuyu and other local tribes worshipped God residing in snow-topped, cloud-wrapped Mount Kenya, the master of the rains, who presides over these privileged lands.

Northwards of Nakuru, the Great Rift Valley descends as it spreads towards Lake Turkana and Ethiopia. The Laikipia Escarpment falls abruptly in gaunt cliffs to Lake Bogoria where the blue salty waters are painted by pink skeins of flamingoes.

Immediately to the north, the oval shape of freshwater Lake Baringo sprawls. Between the two, fed by rivers from the heights, there are swamps filled with tall green reeds surrounded by yellow-boled fever trees and a small tribe live there from scattered cultivation, stock-rearing and fishing. They are the Njemps people who used to farm this naturally-watered place with intensive methods and irrigation more than a hundred years ago.

The Tugen Hills, where hominid fossils older than the *Australopithecine*s were found, loom above Lake Baringo to the west and along the line of the escarpment there is a chaotic mix of natural bush and the endlessly differing ridges and peaked knolls of the underlying land. Nowadays, on the stony eroded lower slopes of the valley, scatterings of goats denude the thornbushes and croton scrub and on those jumbled flanks there are a few far-scattered homesteads. Each homestead has two or three simple huts, a small goat kraal and a patch of maize and vegetables. Within the thick bush goats and a few stubby cattle move about. On the plateau above, where the forest has been cleared, there is a patchwork of *shambas* where maize, sugarcane, vegetables, bananas and pawpaws grow richly.

The landscape may have been the same 3,500 years ago. The maize gardens would have been sorghum and millet, but looking down from the top it would be easy for one's imagination to ignore the detail of difference in the cultivated grains. Close examination of the goats and cattle would have shown how the species had diverged over long centuries of breeding. The fields were cultivated now by iron hoes rather than stone, but none of those superficial changes has relevance to the appearance of what was a picture-story showing how the land may have been after the first Late Stone Age farmers pushed in.

Beyond the modern town of Kabernet there is the deep, narrow Kerio Valley; an arm of the Rift which there splits the highlands which begin their steady slope downwards into Uganda and the huge basin of Lake Victoria. The shores of Victoria are 1,200m above sea level and from them the land gently ascends to the mountains averaging 2,250m on the rims of both arms of the Great Rift. Volcanoes tower over the highlands of both arms. To the westward of Victoria are the Ruwenzories, the fabled 'Mountains of the Moon' with glaciers and snow clad peaks rising to over 4,800m; to the east there are Mounts Kenya (5,199 m) and Kilimanjaro (5,892 m). In this vast and luxurious land lying between the two arms of the Great Rift Valley and encompassing Lake Victoria, which generates its own weather system, there was a great vortex of African history.

It is a huge area, rich and well watered and, before European colonial powers drew arbitrary lines on the map, it had the unity of geographic integrity. There are several modern states covered by it: north-western Tanzania, Burundi, Rwanda, a large slice of the Congo, southern Uganda and western Kenya. Lacking anything better, scientists refer to it as the Interlacustrine Zone. This vast geographically integrated Eden is the size of France or the Iberian Peninsular.

*   *

Until recently, a superficial view of African history had a giant gap, as if the 'darkest continent' was not only a geographical term but also an historical one. There were considerations of the emergence of mankind and a picture of Early Stone Age hunters living a primitive, 'savage' existence on the great savannah plains surrounded by a myriad of wild game. There are the marvels of ancient Egypt and the stereotyped images culled from the Bible and stories of the pyramids and the empires of the Pharaohs.

And then there is the giant gap until a vision of Arab and European adventurers appears, striding onto the scene always with a rifle or embellished muzzle-loader crooked under their arms, exploiting simple natives who are led, semi-naked, in long safari columns carrying loads on their heads. Those pictures are placed on a background of the jungle along the Bight of Benin, or on the veld of eastern and southern Africa. The connection between the three caricatures; primitive mankind, ancient Egypt and modern colonialism was always missing. It was easy to assume that Egypt had no real part to play in Africa, being oriented to the Middle-East, and that nothing much separated the primitive, Stone Age 'savages' from the people first met by European or Arab travellers. It was not only Europeans that held this view as a reading of the medieval Arab geographers shows.

In the centre of 'darkest Africa', great events occurred which fill in the gap of history. But, it is only in recent years that the outlines of a picture of the transformation of hunter-gathering, Late

Stone Age society of 5,000 years ago into the highly-organised tribal and national structures of the Late Iron Age have been drawn by archaeology. Sub-Sahara Africa never slumbered in slothful apathy. There were always influences from the outside and change from within. The climate never ceased to stimulate. What confuses a superficial view is that there do not seem to be pictures of great dynastic empires and astounding leaps in technical innovation.

In the Interlacustrine Zone surrounding Lake Victoria, in the millennia since the end of the last ice-age, people emerged from the forests and met others who moved directly south from the upper Nile as the Rift Valley became flooded, rivers ran with greater power and the dry savannah greened with forest and lush grasslands. They both probably confronted expanding bands of nomadic herders from Ethiopia. Here *Bos taurus* Saharan cattle met *Bos indicus* from the Horn of Africa and the *sanga* breeds resulted. Agricultural techniques filtered down from Egypt, and later mining and smelting specialist clans found iron ores and spread their magic. Sahel sorghum cereals were hybridised and so were the fruits of the woodlands. The migration routes are clear enough when one studies the geography.

Change came with the new technology and the trappings of property, derived from the previous 5,000 years of social and economic evolution in northern Africa and the Middle East since the end of the last ice age. Populations of agriculturalists increased and the hunter-gatherers who were living there were absorbed or pushed into areas where agriculture was not profitable. The Hadza people of Tanzania were remnants of the indigenous hunter-gatherers of eastern Africa, click-speakers related culturally to the San-Bushmen of southern Africa who shared the same fate a thousand and more years later. Similarly, the Dorobo and Njemps of the Rift around Lake Baringo and the Cherangani Hills may also have had these ancient genetic links, obscured now by merging with agriculturalists and their cultures.

In West Africa where population compression was endemic with the extending Sahara, complex villages developed into fortified towns and walled cities with complex social hierarchies. Although probably relatively modern, the extensive earthen walls and ditch system of the Eredo complex, 100 miles in length, have been traced deep in the Nigerian rainforest. In the Interlacustrine Zone around Lake Victoria, evidence of similar societies has been found. There are remains of defensive villages with earthen walls and perimeter ditches in Uganda and western Kenya. They resemble pre-Roman hill-forts that are scattered over England.

Population pressures must have become acute from time to time in this Eden once all the easily utilised lands were occupied. The availability of new land for cultivation was limited by the ability of the inhabitants to clear the forest. When there was sustained population pressure, insoluble from not chopping down trees fast enough or by negotiation and temporary clientship, warfare or the threat of it had to result.

In the millennium after 3,500 years ago, this Eden began to fill and tribal wars of varying degrees of severity had to begin. If the newly-learned strategy of warfare did not resolve the problem, the ancient way to alleviation was to move. The problem of refugees fleeing population pressures and territorial conflict emerged in Africa in rich lands like the Interlacustrine Zone at that time, and their effects were later felt as far as the southern Indian Ocean shores. It is not coincidence that most of the worst excesses and savagery of modern post-independence civil wars in Africa have occurred in this particular area and in the West African forested lands.

Some farmers had to move on. In West Africa this was not always possible and urban society developed to highly sophisticated levels, but in Central Africa there were possible routes outwards. They labouriously followed the rivers down to the ocean where they found hills eternally greened by the monsoon, following the same routes which *Australopithecus* and the early *Homos* used back-and-forth between the seaside and the Great Rift Valley.

Between the highlands of both the Interlacustrine Zone to the west and Ethiopia to the north and the monsoon-laved coast there is dry savannah which is parched and uninhabitable for much of the year away from those few rivers which provide the ancient highways.

In Ethiopia the gradual reduction in rainfall created pressures there too. Late Stone Age people together with their agricultural technology and social culture moved south along the seashore of the Indian Ocean. Their populations were also growing and they needed room which the coastlands moistened by the double monsoons provided. There is increasing evidence that they were in advance of the trickles down from the Interlacustrine Zone, but the two movements were generally coincidental.

The problem still to be resolved is the identification of the people who carried the economic and cultural change, or who were involved with its diffusion so early. Professor Felix Chami of the University of Dar-es-Salaam reckons that it was Negro people of the Bantu-language group who were the original inhabitants of the whole region at 5,000 BP. This, of course, is open to dispute and argument, and conflicts with my own view which is that the San, or San-related people, were the natural inhabitants of eastern African savannahs and that Negroid people of Nilotic, Afro-Asiatic-Ethiopian and Bantu cultural or racial sub-divisions all expanded from their heartlands over eastern African at this time of general population growth, pushing the San hunter-gatherers into lands unsuitable for animal husbandry. Chami does not attempt to detail the movement of people or their origins previous to 5,000 BP, and thus it is not possible to examine his reasoning.

Whatever the outcome of further investigation of the time before about 5,000 BP, it can be sure that Negroid people began to trickle southwards along the Rift escarpments past Lake Tanganyika to yet another vortex in the highlands of southern Tanzania, north-eastern Zambia and around Lake Malawi. There were fly-free connections and corridors which could be patiently explored for the movement of cattle and to lands suitable for cultivation. Here they met those who had migrated southwards, west of the Congo, and moved into the savannah lands of today's northern Angola, southern Congo and Zambia. Neolithic agriculture and then the new Iron Age economy and industry began to filter into the vastness of southern Africa, wherever geography suited. The San-Bushmen hunter-gatherers, descended directly from the ancient Core-People, whose ancestors had roamed without competition for millennia, had their lands alienated. They were absorbed or driven to mountains and semi-desert.

*   *

Apart from building defensive villages, people practised specialised intensive agriculture when climatic or population pressures required it. Good land was often at a premium when people were in surplus or climate inexorably brought change. This is always the cause of turmoil in agricultural society. When migration is not practical, another way out of the problem is innovation and Africans frequently used this technique. History describes the particular example of irrigation used by the Njemps between Lakes Baringo and Bogoria. I believe this example can be extrapolated backwards in time and place.

The surge in agricultural technology by the Njemps in historical time was stimulated by Arab and Swahili ivory-collecting and slaving caravans which used that well-watered and pleasant place as a stop-over depot and trading centre.

David M. Anderson in a paper published in *Azania* in 1989 wrote:

> It can be argued that at Lake Baringo - and perhaps at other irrigation sites also - in contrast to the norm in much of nineteenth century Africa, land and not labour was the major constraint on production. ... In a situation where additional labour could be found with relative ease, this [the operation and extension of irrigation by organised communal labour] was a less onerous proposition than it may have been elsewhere in East Africa. But it is reasonable to suggest that farmers would first seek to increase production within the existing irrigated area by making modifications to their agricultural practices, as this could be accomplished at the level of the household and without involving the consent and participation of the wider community.

The reason why a simple community of farmers, settled in a particular well-watered area surrounded by dry thornbush, should go for irrigation, firstly by individual innovation and then later by communal work organised by community leaders is simple. They had plenty of people and limited land and to satisfy the needs of an underutilised population they decided to create a surplus to barter with trading caravans coming up from the coastal ports. It was an external stimulus that caused the extra effort and when the caravans ceased, the system declined.

It could be argued that Arab or Swahili caravan leaders suggested irrigation and organisation to them and that the intensive cultivation there was not indigenous. But Anderson points out that it was recorded in 1888 that Swahili traders suggested maize as a crop to the Njemps at Baringo and gave them seeds, but that the farmers did not pursue it. The traditional crops were sorghum and millet and these were understood from the mists of time, and I am persuaded that irrigation was also understood by long-gone generations.

The agricultural technologies of the Nile of 5,000 years ago were carried southwards and maintained for many centuries. I do not think it was difficult for the Njemps people of Baringo to revive intensive cultivation under external stimulus. If their knowledge was rusty, experiment sharpened it up. And if a passing Arab suggested leading water in a canal from the river to flood a field, his suggestion took root because it was part of their culture in folk-memory. Increased harvests of their traditional grains proved the value of the extra labour and the risk of introducing new, untried crops was doubtless considered unnecessary.

Somewhat further to the north up the Kerio Valley, on the flanks of the Elgeyo Escarpment, there has been extensive agricultural terracing and in many places on the healthy highlands, stone-lined calf and goat byres like those at Hyrax Hill were built: the so-called Sirikwa Holes. Both the rough terracing on the flanks of the Elgeyo Escarpment and the widespread Sirikwa Holes are from the Late Iron Age, during the last millennium, but they are further clear evidence of two traditions of pre-historic African agriculture: intensive cultivation and mixed farming with its associated cattle-cult.

A.T. Grove and J.E.G. Sutton in an article in the special edition of *Azania* in 1989 devoted to agricultural technology commented on the historical conclusions that may be drawn from terracing. Evidence of terracing and irrigation right across the Sahel of West Africa, in the hills of southern Sudan, extensively in Ethiopia, both arms of the central Great Rift Valley, southern Tanzania and eastern Zimbabwe, has been written about often enough. Grand theories were proposed and the politics of African historiography inflamed.

Although discredited now, it has been argued that agricultural terraces and stone building was evidence of colonisation of eastern Africa by various external megalithic 'civilisations' long before the appearance of Europeans. A strenuous argument was put forward by R. Gayre of Gayre in several publications. Gayre drew much inspiration from G.P. Murdock, an influential anthropologist in the 1950s, and in his book, *The Origin of the Zimbabwean Civilisation* (1972), sought to link kingdoms of the Yemen directly to Zimbabwe and proposed that stone building technique and intensive agriculture with irrigation was introduced by Sabaean and other Arab colonists and settlers.

Gayre and others have been often unfairly dismissed as having a deliberate racist bias (students at Edinburgh University rioted when he tried to address them), but dismissing misguided and already obsolete ideas of Semitic empires in Africa should not erase proper consideration of trade and the exchange of knowledge between the Nile civilisations, southern Arabia, the Horn of Africa, southern Sudan and the central Great Rift Valley system with its verdant Interlacustrine Zone. It is sadly true that those who are most virulent in their condemnation of bias are often the most biassed and intolerant critics of all.

Egyptians and Nilotic people in the Sudan built great stone temples and monuments as well as extensive cities of square mudbrick houses. They had intensive cultivation and irrigation techniques organised and taxed by a vast bureaucracy. However diluted, that knowledge and culture spread over the Sahara and along the Sahel. There were farmers spreading southwards along both limbs of the Great Rift Valley in East Africa. People were moving south from the Ethiopian highlands. By then, this was becoming part of African culture and there seems no reason for Africans to find its permeation of eastern and southern Africa objectionable.

Kingdoms in the Yemen were powerful and prosperous. Wealth was created by trading in frankincense directly with Egypt, kingdoms of the upper Nile in the Sudan, the Middle East and India and acting as carriers and middlemen of the Indian Ocean trading system. A food surplus was needed to support the trading cities and ports. On the southern Arabian highlands where the monsoon rains fell, a dam with sluice gates feeding irrigated lands was built. On the sea-facing escarpments, vast arrays of terraces were built to hold moisture and conserve the soil. Those terraces are still in use today.

Frankincense was also grown in the Horn of Africa whose people were also part of that trading system and despite the occasional dynastic dispute and wars that raged for a while, Ethiopia and southern Arabia maintained a close economic relationship with an interchange of people and ideas until the rise of Islam in the 7th century AD. Stone buildings and terraces were common in Ethiopia.

A.T. Grove and J.E.G. Sutton in *Agricultural Terracing South of the Sahara* (1989) wrote:

> Now that diffusionist theories, let alone notions of 'higher' cultures, have been in retreat for some time - being not so much proved wrong as just considered irrelevant to modern social sciences including archaeological studies in Africa - the question of testing anew these old attempts to survey and map terracing continent-wide may barely seem worthwhile. However, with terracing coming to be recognised as a specialised technique, albeit usually a locally evolved and a peculiar one, commonly integrated with other specialised agricultural and socio-cultural features, there is a renewed interest in the value of comparisons, not for theories of historical diffusion but simply for helping to understand the workings of those systems.

There can never be any doubt about the emergence of many advanced techniques and all the social complexities of civilisation, led by agriculture, along the Nile, which spread across the Sahara and thence southwards. All of the people who moved south, gradually pushing aside or absorbing the hunter-gatherers of eastern and southern Africa in the last 5,000 years were agriculturalists; whether cultivators, mixed farmers or herders. They had the inherited knowledge, they changed their roles as geography dictated and they exchanged ideas and products with itinerant traders, neighbours, clients and overlords as the centuries rolled. Those who found lands that suited the particular economy that had become traditional to them, and held on to them, became specialists for greater or lesser time. The new agricultural technology moved southwards, ousting hunting and gathering, and there is no question about that. There can be no doubt that communities of hunter-gatherers were converted to agriculture by clientship or observation and emulation, and this is important to academic controversy about diffusion versus migration.

The contrasting or interacting mechanisms of the transfer or geographical movement of culture by diffusion or migration has always been fascinating. Archaeologists and historical anthropologists increasingly seek evidence for the prominence of one or the other as archaeological exploration of Africa gradually progresses, but this is difficult. It is a complex issue and in my view is impossible to resolve in many African environments. Even where the most intense attention has been focussed, as at Great Zimbabwe, all questions are not finally answered. I firmly believe that the spread of culture takes place in both ways and in a combination of them at different times with different people and circumstances. It is easy to forget that we are considering 5,000 years. My fascination lies in trying to understand when which culture moved and who was involved.

If for whatever reason of pressures by climate or other people, or simply for economic advantage or convenience, the concept of irrigation or terracing was revived from time to time in suitable places, then this was the natural revival of a specialist traditional technique imprinted in them. Where the land was rocky, stone is an obvious material with which to build domestic structures and agricultural terraces. Terracing and irrigation, with associated stone-walled homesteads and cattle kraals, were within the repertoire of their cultural inheritance, developed over previous centuries, and if ecological change or migration later made terracing unnecessary or inappropriate for continued prosperity they abandoned it. Many irrigation projects and associated terracing in Africa had been abandoned and disused when European explorers discovered and wrote about them in the 19th century.

I believe that when behavioural evolutionary jumps in technique or organisation have been taken and there has been enough time for them to become imprinted, then they are part of communal memory and tribal tradition. Detailed knowledge may be lost, but the concept cannot be expunged. Climate or forced migration caused people to stop farming or herding from time to time and revert to the ancient ways of fishing, hunting and gathering in order to survive. Different people in Africa lost their cattle for generations, but no great revolution was necessary to begin keeping them again when it was possible. Maybe style and method changed with changed circumstances, but the concept did not have to be re-invented: similarly with techniques of cultivation.

*   *

There is another important pillar of African agriculture: the exploitation of exotic crops. The Interlacustrine Zone of the central Rift Valley system was particularly rich in people and the variety of healthy lands for this to occur.

Many exotic plants were introduced to Africa over the last 5,000 years. When early civilisation burgeoned, it was in Asia that many plants were domesticated. In the sub-tropical zones of the Middle East and India and tropical south-east Asia and Indonesia cultivators spent several thousand years developing hybrids which were later introduced, together with farming technique, into sub-Saharan Africa. The imported strains did well enough so Africans did not have the need to spend a thousand years hybridising indigenous plants.

As sparsely-spread gatherers, sub-Saharan Africans were satisfied with their own continent's fruits, roots and seeds, but after the jump to agriculture the crops which have been most successful were developed elsewhere. Over the millennia they have been introduced to Africa, starting with Middle East grains, vines and legumes and accompanied by sheep, goats, pigs, poultry and cattle. When intercontinental trade expanded powerfully from about 5,000 years ago, especially between India, the Middle East and the Mediterranean, via Arabia and north-east Africa, exotic foods came with the traders. Domestic fowls came from northern India. Coconut palms, peppers, sugar-cane, hybridised yams, rice and bananas are particular examples from the tropical Far East, either directly from Indonesia by sea, or via India and Arabia. Citrus fruits came down the Nile and by sea from the Mediterranean.

In recent centuries, varied poultry and mangoes from India were introduced by Arab and Indian seatraders. The Portuguese brought maize, cassava-manioc, potatoes, tomatoes and avocados from the Americas and spices from the East Indies.

Bananas are indigenous to central and west African forests and there is evidence to support some selective domestication of native species. But the bananas which became the dominating staple of the Lake Victoria region and an important supplement throughout the tropics are Indian, Malaysian and Indonesian species. Speke and Grant were the first Europeans to explore the western side of Lake Victoria and they were amazed by the high standard of banana cultivation and the many uses to which the leaves and fruit were put. The leaves were used for thatching, utensils and clothes; the fruit was eaten raw, boiled, roasted, dried, powdered and fermented.

C.C. Wrigley prepared a useful summary of this fruit in his paper, *Bananas in Buganda* (1989). In the introduction he wrote:

> It is pleasant to write about an agricultural system in Africa that was as nearly as possible trouble-free - that is to say, one which reliably provided an adequate supply of food to a fairly dense population with modest inputs of land and labour. Famine in the strict sense has never occurred in Buganda, and rare occasions of widespread hunger have had political, not natural, causes.

In historical times around Lake Victoria the banana was the staple food supplemented by beans and other vegetables and protein was supplied by domestic fowls and fish. Fish was traded fresh by the shore and dried inland. There were goats and some cattle, but no cattle-cult. Beef was eaten as a luxury and raided or bartered from people who lived in drier, disease-free country further afield. The property of wealthy lakeside tribes was their banana groves and fishing rights in the rivers and on the lakes.

Asiatic bananas first came to Africa as portable food in ships from India, either directly or via the coastal trading nations of Arabia, and the magnificent sailing canoes of Indonesia. Indian sailors were trading with Arabia and African Red Sea ports before 2,000 years ago and Indonesians colonised Madagascar and the northern Mozambican coast at about the same time and later in two distinct waves. Because the Indonesian presence on the eastern African coast is overshadowed by later events, and their principal colonies were on Madagascar which is a 'forgotten' land, their influence is often overlooked.

Wrigley sums up his view of the introduction of bananas to the Interlacustrine Zone:

> First, probably some time in the first millennium AD, *balbisiana* hybrids were brought from India to the east coast of Africa. Being somewhat better adapted to dry seasons than the pure *acuminatas* of the equatorial forest, they were able to make their way inland to the Lakes region and beyond. .... Later, *acuminata* forms entered the

southern part of East Africa and followed the generally well-watered '*tooke* corridor' to the Lakes region.

This corridor is the migration track of bananas (locally called *tooke*) from the ocean, where Indonesian immigrants and other seatraders had direct access, up the Rovuma River, through northern Malawi and western Tanzania to Lakes Tanganyika and Victoria. It is one of the few migration and trading routes between the human vortex around the Interlacustrine Zone, the central highlands and the Indian Ocean. The 'tooke corridor' was used extensively as a road to the ocean by Zanzibari slave traders who led their coffles down from Nkhota-Nkhota in Malawi, but it is unsuitable for people with cattle because large portions of those lands were infested with tsetse-fly.

Healthy trading routes for cattle in the dry season ran along rivers further north passing through savannah grassland or desert. The Athi and Tsavo tributaries of the Galana run respectively from highlands around Mounts Kenya and Kilimanjaro to the sea at Malindi. The Tana rises in highlands north of Mount Kenya and reaches the sea in flat swampy country not far to the north of Malindi. Both these river roads have provided communication links far back into the mists of time, and during the last 5,000 years were probably the only connections for herders or traders with vulnerable cattle.

Other connections were from the Somali coast (the classical 'Strands of Azania') and the Horn of Africa to interior highlands through the desert along the seasonal Uaso Nyiro which joined the Juba [Giuba] at Kisimayo in wetter times. The Juba and Shebele provided a route to the Ethiopian highlands which are transversed by the Great Rift Valley. Cushitic agriculturalists, herders and traders moved southwards from Ethiopia using these rivers to the coast, and centrally along the Rift Valley to the Kenya highlands.

Trade and ideas did not move only in one direction, they spread everywhere in this sophisticated environment reaching from the great classical civilisations of the Middle East right into the heart of Africa. We tend to view society before the Industrial Revolution as if it were slow, sluggish and wrapped in fictional mythology and fantastic religion. It was nothing of the kind.

*   *

Increasing evidence of trade from the Indian Ocean coasts, along these routes, to the southern Nile and thence to the Sahel belt and Egypt is being considered. Felix Chami pursued this following his excavations in Tanzania, particularly in the Rufiji delta and on Zanzibar. If it is correct to assume that technique moved southwards from the Nile region, then it is equally sound to consider that knowledge and imports were carried by traders who chose an overland route from the Indian Ocean to the Sahara and Mediterranean. It is notable that archaeologists have found evidence of a diffusion of culture from the southeast into the southern Sahara during the millennia when the Sahara became savannah, and was being re-occupied by people after about 9,000 years ago until the drying cycle recurred 3,500 years ago when people moved outwards again.

Culture and technique moved in both directions; agriculture, metallurgy and urban pastoralism from the Middle East into Africa and gentler forms characterised by pottery, food cultivars and style from eastern Africa to the northward.

In the last few years the evidence for the occupation of central and eastern Africa by Late Stone Age farmers, whether accompanied by cattle or sheep, has been steadily increasing. Archaeological discoveries in the Interlacustrine Zone and along the Tanzanian coast show that Late Stone Age people were practising agriculture well in advance of the Iron Age migrations of people with Urewe pottery, presumed to be Bantu-speaking.

Professor Felix Chami has published results of his pottery finds and fossils which show that Late Stone Age agriculturalists were living along the Tanzanian coastal monsoon belt before the dates previously assumed, at about the end of the first millennium BC, categorised by Kwale pottery types from the important Pemba River site near Mombasa. His latest discoveries indicate that Late Stone Age culture was uniformly widespread in the first millennium BC.

Professor Paul Sinclair of Uppsala University who has spent many years investigating the archaeology of East Africa published this on the Internet:

> A number of other caves have been located by Chami on Zanzibar and on Juani Island near Mafia 25km off the coast of Tanzania. Preliminary investigations by Chami

have resulted in finds of similar quartz and limestone stone tool assemblages to those reported from Kuumbi as well as bones of antelope and domestic chicken. Finds of South Asian Ceramics and Beads made from glass and carnelian stones and pottery provide evidence for participation of these communities in the trade networks of the Western Indian Ocean from c. 400 BC onwards and these findings are the focus of joint research led by Prof Felix Chami of the Department of History University of Dar es Salaam, Dr Abdurahman Juma of the Zanzibar Antiquities Service and Prof Paul Sinclair Uppsala University.

In *The Unity of Ancient African History 3000BC - 500AD* (2006), Chami writes:

> ... The dates [of digs on Mafia and Zanzibar islands] are of the 1st millennium BC. Of equal importance is research on the mainland coast and interior that established Neolithic sites, as reported elsewhere: Rufiji Delta region, Lindi and Mtwara regions, the Rufiji catchment, and the Lake Victoria region. On going research by Kwekason in Lindi/Mtwara coast has now discovered three more Neolithic sites beginning with Khartoum/Kansyore tradition.

Chami continues, after describing other similar sites:

> In the examination of all the pottery collected from those sites discovered by the research summarized above, it has now been established that all of traditions of the Nile Valley Neolithic are present on the coast and islands of East Africa. Kansyore/Khartoum pottery is now known from Kilwa Island and Nderit pottery from Mtwara and Kilwa Island. The most widespread is Narosura pottery, found in the dinosaur site of Tendaguru and the Rufiji Delta region. Narosura settlement at Kilwa Island also seem to have been large. Pottery related to some Nile and Rift Valley ceramics, including Narosora and Ileret/Khartoum Neolithic, are also found in a site in the catchments interior of Rufiji (150 km from the coast) and in the Mchaga Cave in Zanzibar. [I have omitted his detailed references.]

Undoubtedly, the most important new information in Felix Chami's book is his note of excavations on Zanzibar. I quote:

> Since then a new research in another cave in Zanzibar, Kuumbi, has found more chicken bones in a large quantity in the Neolithic context, now in association with cattle bones (Bos taurus). Their contexts have now been dated to between 3300 BC to 0 BC.

This site has also provided evidence of early seatrading activity along this coast, earlier than had been previously established. Indian pottery from the first millennium BC had been earlier identified by Chami at the important Machaga Cave site on Mafia Island.

*   *

The Early Iron Age south of the Sahara is accepted to have been established in the first millennium BC. It is associated with speculative variants of Urewe pottery from as far west as the Niger River at Jenne across the Sahel to the upper Nile in the Sudan. Urewe pottery is most closely defined as having its nascence in the Interlacustrine Zone and it has been usually accepted as being related to the Bantu-speaking culture, carried down to the coast with iron technology at around the end of the first millennium BC. The Pemba River site at Kwale (Kwale pottery) has been the defining archaeology. Radiocarbon dating of many sites in the lands around Lake Victoria show that Urewe tradition pottery, and iron-working locations, proves occupation by people with those cultures from at least 3,000 years ago. There are several dates from the 2nd millennium BC as detailed by Marie-Claude van Grunderbeek in *Azania XXVII* (1992).

It is entirely likely that the smelting of iron was commenced along the Sahel in advance of its general production in the Nile civilisation. Provided that the general knowledge of the use of intense heat and reduction agents such as charcoal in the 'melting' of ore to create metals is available, the discovery of iron smelting is inevitable. The economy of Sub-Sahara Africa along the Sahel and in the Interlacustrine Zone, in the first millennium BC is closely linked to the beginning of the Iron Age, the use of iron tools in clearing forest for agriculture and thence to the rapid increase in populations, leading to migration or territorial conflict.

Felix Chami demonstrated by his work in the last decade that simple models of transfer of culture and economic activity in East Africa have to be reassessed to some extent, sharpened as they have been by his fascinating new archaeological sites.

Pottery has shown that by 2,000 years ago farmers with Urewe wares and iron-working technology had made their way down from the vortex of the Interlacustrine Zone across the dry savannah using the ancient river routes. The coastal strip is a paradise for cultivating farmers because the monsoons bring regular and almost continuous rainfall. Mombasa has an average of more than two inches of rain every month of the year, except January and February, with a peak of about fifteen inches in May. On either side of the Kenya coast there are breaks in the rains: further south at Zanzibar there is a dry season from June to November and north at Mogadiscio, in Somalia, it is dry from September to April.

It is not surprising, therefore, that the Kwale district adjoining the coast just south of Mombasa is famous for Early Iron Age pottery made by farmers about 100 AD who may have migrated there from the interior. Kwale pottery has stylistic links with Urewe made in the Interlacustrine Zone. There may not have been many of these migrants at first, and the idea of a determined national movement is not sensible. There were ripples of movement, far apart, when temporary population or climatic pressures in the highlands persuaded adventurers to seek new pastures.

There were people already there and Chami has shown conclusively that Negroid Late Stone Age people with long-established links to Indian Ocean seatrading had been well-established for several thousand years. Africa has never had vacuums, not since *Homo erectus*. Farmers and maybe Afro-Asiatic Cushitic-speaking herders from the north, and coastal fishermen, knew the area. Rhapta was an established kingdom around the Rufiji delta 2,000 years ago. These are important and fascinating themes which must be explored in later chapters.

In 1987, I observed a small *shamba* homestead in the Pemba valley behind Kwale where the famous Iron Age pottery finds were made. There were a few round wattle-and-daub huts with grass thatch roofs, a raised grain store, a patch of cereal, pumpkins on their vines, bananas, a paw-paw and a mango tree. A bleating goat was tethered on a string. A naked girl-child with coloured beads strung about her belly ran out to wave. The only obvious evidence of the passage of two millennia was the printed *kanga* cloth drying on a thornbush and an enamelled metal bowl, substituting for Kwale pottery, lying by the three-stone fireplace with its wisp of aromatic woodsmoke. I searched for that *shamba* in 1991, but it had gone in the intervening four years. That *shamba* could have been duplicated anywhere from the Kenya coast to the distant mountain valleys of the Ciskei in South Africa.

THREE : *THE KHOISAN*
The mysterious spread of nomadic herding to the far south.

The Late Stone Age aboriginal people of southern Africa are today usually termed the Khoisan which is an artificial name of convenience.

The ancestors of the Khoisan of southern Africa split and consolidated in two quite clearly defined socio-economic and language divisions. There were those who kept to hunter-gathering and lived in small, politically independent bands and are defined as San, commonly called Bushmen. Others adopted a pastoral herding life with exotic sheep possibly as early as 2,500 years ago, and cattle later. They developed loose clan federations and territorial concepts. Some of these divergent people were known by themselves in historical times as the Khoi-Khoi: 'men of men'. Some of these Khoi-Khoi called their hunter-gathering brothers, San. The race as a whole therefore became defined by historians and anthropologists as the Khoisan.

The pastoral branch of the Khoisan, the sheep and cattle herders, were defined separately as Hottentots in formal literature since 1677, but this term, which was first used at least since the early 17th century by Europeans, became politically incorrect because it was used in a pejorative racist way when referring to the mixed-race 'Cape Coloured' peoples of South Africa. I like the traditional name, Hottentot, and it could have retained an honourable status, but it was misused and degenerated. Therefore, I have not used that word, unless it is part of the context, and use an accepted term, Khoi or Khoi-Khoi.

Recently, other names have gained popularity in South Africa which are Khoe, Khoekhoe, or Khoe-Khoe. Quena, or Kwena, has also been suggested as another collective word for them. In

South Africa in recent years there has been an increasing tendency to change names for places and people which is confusing and sometimes has political motives. None of the indigenous people of southern Africa had their own names for their race or culture. Always people called themselves by the name of their clan, or clan chieftain, and this name changed as dynasties changed. Since they were not literate, the spelling of names changed with modern political or academic fashion too; another source of confusion. The name used for a language or dialect is often applied to a whole culture, or quite erroneously to a race. Khoe, for example, has become a new name for the language spoken by the modern Nama people of Namibia and northwest South Africa and increasingly is applied to the whole culture and, by association, with the ancient Khoisan race.

Pastoral herding was adopted by some Khoisan and those who took this jump began developing a significantly different culture about two thousand years ago in South Africa. How they acquired domestic sheep and cattle is controversial. Whether the same jump occurred elsewhere amongst Khoisan people in central and eastern Africa and when that might have happened is not known and is also controversial.

There were no distinct Khoisan left outside South Africa, Botswana, Namibia, Angola, and a few pockets in Zimbabwe by the 19th century when European explorers wrote about those parts. But there was much speculation about their origins and the racial composition. The first informed European observers in southern Africa were very clear about the distinct differences between the Khoisan of the Cape and the Negro peoples further to the east and north and late 19th century writers and academics were not inhibited by hints of political correctness to voice some theories of origin which are quite astonishing today in their wild speculations. It was fashionable to ascribe any kind of superior culture to the classical civilisations with a sharp dividing line between 'civilised' and 'savage' people. This does not mean, of course, that one should dismiss their valuable firsthand observations and experience of indigenous people.

*

I travelled in the Kalahari in 1962, in 1972 and several times in the 1980s and 1990s. Five minutes out of Tshabong in Botswana on the southern rim of the Kalahari, one was deep into the wilderness. As soon as you left behind a few shacks and the footpaths to them that were on the outskirts, there was no evidence of people but the narrow ribbon of churned sand that led northwards. For the first few miles you passed over undulating ridges of bright orange-red sand dunes and yellow clay had been laid to harden the surface of the track. Thereafter, low scrub covered the dunes which gradually disappeared to be replaced by flat land which rose and fell over distances of several miles in a sequence of shallow swells. The yellow sand track stretched ahead through monotonous, rough olive-green scrub.

Bosobogolo Pan was a good place to camp for some days and absorb the Kalahari wilderness. You may have found a small herd of springbok and some scattered wildebeest and gemsbok, the South African oryx, looking up and moving nervously away from your vehicle. At night, out on the bare flat of the pan, there was the occasional faint scuffling noise of antelope moving and jackals yelping to each other. When a torch, or flashlight, was shone across the void, eyes reflected back. Usually there was the whooping of hyenas and the distant, grunting roar of a lion. At dawn, kites and kestrels wheeled about looking for their breakfasts. Koraans set up their distinctive staccato call, echoing each other, and several species of birds of the dry bushveld played music. The resident gemsbok antelopes moved about slowly, carrying their long savage horns with dignity.

Bat-eared foxes and black-backed jackals came out in the evening and patrolled at night. The foxes are attractive animals and their diet is similar to baboons, or Stone Age people for that matter, consisting of small animals like lizards and whatever insects they can catch in season, chicks and birds' eggs, tubers and roots. They mate for life and live in nests in the ground during the heat of the day. Unlike baboons, they can survive without water, gaining enough liquid from their food.

In the cool of the approaching evening springbok were lively, springing with straight legs, *pronking* Afrikaners call it, then racing about in a flurry of dust, playing together and the males practising the lines of dominance. When some springbok ran too close to a group of grazing gemsbok, the bigger antelopes also leaped off with a startled jump and spring. Giant eagle owls had their territories and watched for prey from the branches of dead trees. Ostriches wandered onto the pan and found a favourite place where they had a vigorous sand bath, crouching down and shuffling, waving and fluffing their wings and sending up a stream of dust into the breeze. There were many ground-squirrels, mongooses, termites and ants, various beetles, butterflies and spiders. Around the pan's edge there were tiny veld-flowers: blue, pink and yellow. At night moths came to camp lanterns and one could hear nightjars and see bats flitting by.

Considering that there is no surface water except for a few weeks after seasonal rains which often fail, the Kalahari Desert abounded with life. In the Mabuasehube Reserve, nature was in balance, at ease and flourishing: it was not difficult to imagine Stone Age hunter-gathering people living there in some comfort. But they do not live there any longer; the absence of the San-Bushmen is the missing image from an ecological picture of the Mabuasehube Reserve.

The great Sahara Desert must have been exactly similar country 5,000 to 10,000 years ago during the 'Holocene Wet Phase'.

Scientific observers have discovered how water moves up the food chain along with other nutrients in the desert. When the land dried after the short showers of summertime, water was stored by the vegetation. Grasses, vines and bushes were flushed with water in their roots and leaves and antelopes ate them, absorbing liquid. When the leaves dried, hard-skinned fruits, tubers and roots held water for months, and these were eaten.

Not only did the herbivores know about them but so did the San-Bushmen who could also gain enough water from them to survive. Elizabeth Marshall Thomas in *The Harmless People* (1959):

> There are many kinds of wild roots which can be eaten in winter [the dry season], and each is marked among the grass blades by an almost invisible dry thread of a vine. The roots are swollen with liquid by which the plants preserve their life during the drought ...
> ... [Tsama] Melons are eaten as both food and water, their pulp is added to meat which needs liquid for boiling, their seeds are roasted and eaten or ground into powder and used as flour, their rinds serve as mixing-bowls, as containers for small loose objects, as

> cooking-pots with or without the pulp inside, as urine-containers for curing hides, as targets for children's [arrow and spear] shooting practice, as children's drums, as resonators for musical instruments ...
>
> ... We were going particularly to look for *bi*, a fibrous, watery root that is the mainstay of the Bushmen's diet during the hot season when the melons are gone... The bi they find is brought back to the werf [group settlement] before the sun is hot: it is scraped, and the scrapings are squeezed dry. The people drink the juice they squeeze.

In these and similar ways, the San-Bushmen survived, but there was another way. It was through eating the animals they killed in the hunt, because there was much liquid in their meat. A sure way of getting instant liquid from a kill was to quickly cut into the stomach and gut and strain the contents to drink the liquid.

With the catastrophic decline of the antelope populations in the late 20th century through interruption of their migrations by veterinary control fences and severe poaching, the economy of the San-Bushmen was also destroyed. The South African author, Sir Laurens van der Post, in *The Lost World of the Kalahari* (1958) wrote:

> For me always the fact of urgent practical consequences was that the Bushman, unlike any possible predecessors, was a remembered and remembering and living link with human origin in my native land.

A culture that endured for 30,000 years and which was still viable sixty years ago has disappeared forever. Undoubtedly, it was doomed though clientship to Bantu-speaking agriculturalists and Western civilisation. But it could have been a gradual process, stimulated by curiosity and slow integration, suggested by word-of-mouth between scattered people. And there may have been time and opportunity to study them more, learn more and write more and add to our understanding of the universal human psyche before the great separations 80,000 years ago with the out-of-Africa migrations.

Of great importance, the special ability of San-Bushmen to survive socially in small extended family groups in wilderness through a thousand generations was lost forever. Whilst scientists argued over fossil bones and anthropological hypotheses, the actual evidence of Late Stone Age hunter-gathering culture in a balanced African savannah environment slipped away from them. Study of introverted tropical rainforest tribes with violent territorial customs, born of those particular environments, in Amazonia or Papua-New Guinea can never substitute for the lost opportunity of learning more about the San-Bushmen in a savannah land similar to those occupied by forebears 30,000 and many more years ago. A unique ancient culture with its peaceful social mechanisms, beliefs, tension-relieving dances, artistic creativity and integrated communal style was gone. They were forgetting the songs and the dances that were as old as the last ice-age.

In September 1975, I joined a field-study expedition organised by the South-West Africa Scientific Society to observe two large parties of San-Bushmen of the !Kung and GiKwe language groups. One cannot talk about tribes of San-Bushmen, because there is no such organisation, but the central culture diverged in subtle ways in different geographic areas. It was probably the last organised scientific meeting with large numbers of San-Bushmen still living by hunter-gathering and in traditional style.

We were at Andara on the Okavango River at the time of full moon and the two distinct groups of San-Bushmen came separately to a suitable place nearby beside the river where each camped for two or three days, erecting their grass shelters. The first group, who were !Kung, were settled in when we arrived and we were able to spend two days observing them.

A South African government liaison official explained where each group originated. The !Kung had apricot-hued skins, fine-boned slim bodies and wedge-shaped, flat faces with high cheek-bones and came from the Kalahari where it spreads into Namibia. Their culture was attuned to the waterless desert, while the GiKwe were from a mopane-forested riverine environment. The latter were darker in colour, almost black, and had rounded faces and short thickset bodies, reflecting a substantially different nutrition regime. They resembled the Pygmies of the Congo. The superficial somatic differences in appearance between the !Kung and GiKwe people I observed showed that even within several days' march environment-prescribed nutrition can influence anatomy given sufficient time.

A temporary !Kung settlement of grass shelters, and !Kung womenfolk at Andara in 1975.

GiKwe women singing and dancing at Andara, Namibia in 1975.

Author's photos

At night, under the moon, they sang and danced. Both groups seemed to be singing and dancing in the same way, but there must have been subtleties, apart from language. It was enough to be amongst those people, so incredibly far away in technical culture. They sang and danced all night long, moving deep into trance-dance states.

Anthropologist Dr Hans-Joachim Heinz was there. He had gained some notoriety through living with a 'Bush-wife' in the Tswaane area of the Kalahari intermittently over a period of many years. He told stories around the campfire and his book, *Namkwa, Life among the Bushmen* (1978), is an important contribution to the literature of Kalahari San-Bushmen.

After the Portuguese explored the African coast at the end of the 15th century, European commentators describe meeting Khoisan people. For example, Augustin de Beaulieu, sailing with a French fleet consisting of the *Montmorency, Espérance* and *Hermitage* from Honfleur in October 1619, gives a long description of the people they met at the Cape of Good Hope. With the typical arrogance of the period, he was not impressed with their primitiveness:

> The inhabitants of the country towards the point of the Cape are, I believe, the most miserable savages which have been discovered up to now, since they know nothing of sowing or of gear for ploughing or cultivating the soil, nor anything of fishing ... They are of very low stature, especially the women, thin, and seem always to be dying of hunger. They eat certain roots, which are their chief food ...   [and so on]

Despite the repugnance to him of these *strandtlopers* as the Dutch later called them, he goes on to describe them in considerable detail. Another example is Thomas Herbert who was at the Cape of Good Hope in 1627 and whose descriptions parallel de Beaulieu. After the Dutch colonised the Cape in 1652, the literature expands. Perhaps the greatest importance of the early descriptions of Khoisan people at the Cape is the close similarity between them and the remnants of San-Bushmen in the Kalahari three centuries later, transcending obvious differences resulting from different environment.

Of modern books Elizabeth Marshall Thomas' *The Harmless People* (1959) is one I find most valuable. Not only did she describe in careful detail the daily rituals and activities of the Bushman extended-families she knew intimately, but she gained a particular insight into the manners with which they resolved social and psychological problems. Her style and personality seems especially in harmony with the gentle and 'harmless' people she got to know so well.

Prof. Phillip Tobias, who studied the anatomy of San-Bushmen, wrote in 1962:

> More and more of the surviving Kalahari Bushmen are abandoning their pristine ways in favour of the assured food- and water-supply of the farms. "Wild" Bushmen who have had little or no contact with European and other farmers are decreasing in numbers and dispersing. ....
> The changes in diet consequent upon the adoption of pastoral habit are both qualitative and quantitative. Bushmen who work on farms are given an assured supply of cereals by the farmer. ...

Shared with the other contemporary hunter-gatherers of wide spaces, Inuit-Eskimos and Australian Aborigines, perhaps the most extraordinary and important trait of the San-Bushmen was their ability to live peacefully. Not only did they have an imperative to resolve all disputes within their bands or family groups, but they had developed a mechanism to ensure no territorial conflicts. Territorial disputes leading to violence did not happen whatever the environmental pressures. Their acceptance of the dominance of the environment, its spiritual authority, and the need for all people to accommodate each other within both catastrophes and good times was unquestioned.

San-Bushmen lived, in the last centuries, in a variety of environments in southern Africa: the Namib Desert fringes, the Kalahari, the Karoo semi-desert, grassland highveld, malarial savannah bushveld, the high Drakensberg Mountains, the Matopos, deciduous mopane forests, the rich riverine lands of the Okavango and Zambezi systems. Each economic group understood, in the most profound way, the ecology of their ancestral territory. San-Bushmen of the deserts did not migrate en masse to the Okavango Delta to invade and overcome the incumbents. That kind of destructive territorial conflict had been worked out millennia ago.

Another remarkable trait of the San-Bushmen, learned over many millennia, was their ability to limit population. Women are less likely to conceive whilst still suckling and when there is severe hardship, stress or famine, women may cease menstruating and cannot conceive. But San-Bushmen women deliberately avoided conception until their youngest child was capable of long-distance walking. This was controlled by abstinence, withdrawal before ejaculation or most often by mutual sexual enjoyment without penetration. If conception accidentally occurred, the pregnancy was terminated, a barren relative cared for the child or, in the last resort, the child would be allowed to die by neglect. Everywhere, nomads in harsh environments practised similar apparently brutal custom, for the simplest reason; a family group cannot trek if the women have to carry more than one child.

Another reason for exponential population growth is insecurity. Rapid social change and the destruction of old moralities promotes fear of the future and a hopelessness. People also tend to have more children than necessary for group survival when it is believed that they will die from expected disasters or move away forcibly or voluntarily. I consider that these are the principal reasons for population explosion in Africa today where cultures are being changed with bewildering speed under the onslaught of technical civilisation and disruption from severe ethnic and political conflicts. San-Bushmen, in common with hunter-gatherer people of open space elsewhere, valued a stress-free life and this was an important ingredient in cultural stability; in its turn an important ingredient of population stability.

Assistance and sharing during cyclical climatic disaster was practised until the temporary circumstances were over. If some members of different groups, forced together for a while by adverse climate, transferred their allegiance, particularly by marriage, this was normal and provided healthy mixing of genes. In any case, by some mysterious psychic force, neighbouring bands sought each other out, at intervals sometimes years apart, and joined together for days of feasting and endless dancing. At those times, young people made liaisons outside their own extended family groups and healthy mixing took place.

When a series of disasters debilitated a group, they attached themselves to a welcoming stronger band as subservient clients until the natural hierarchy of individuals was gradually worked out and a new structure emerged. The principle of clientship leading to absorption was developed in ancient time in the face of natural disaster or bad luck.

However, a simplistic model of San society based on a few groups in the Kalahari in modern times may be too naive or idealistic. A universal 'sharing' or egalitarian society with an easily defined structure may not always have been possible throughout the many different environmental or geographical zones in which San people lived for many thousand years from the Sahara to the Cape of Good Hope. Indeed, this would be a naive view. Elizabeth Marshall Thomas' descriptions of life amongst Kalahari Bushmen in the 1950s can only be a snapshot of particular groups in a particular harsh environment. Late Stone Age people of the Khoisan race lived everywhere from forest fringes, lush riversides, high well-watered plains, on mountain ranges and in dry savannah and semi-desert. Those remnants who have been most closely observed in the 20th century have been within or close to the Kalahari.

Karim Sadr of the University of Botswana in his paper *The First Herders at the Cape of Good Hope* (1998), although writing about early Khoi herders, summarised the several egalitarian aspects of San-Bushman society. They range from sharing being restricted to hunted meat, a way of equalising different hunting skills, establishing social relationships, that some plant foods are shared or not shared and so on. He quotes anthropologists who studied the San-Bushmen with greater breadth and rigour than Elizabeth Marshall Thomas or Laurens van der Post. Sadr wrote:

> It seems that sharing can have very different causes and ramifications in different hunter-gatherer societies and in different periods. Sharing is resorted to when it enhances the hunter-gatherers' long-term success in obtaining food, but is not blindly adhered to. ... Thus, when a group's particular ethos of sharing becomes non-adaptive for whatever reason, it can be ditched.

It could be pointed out that this is also quite obvious and that variation of behaviour must occur with different geography, and one can imagine circumstances which gave rise to variations. However, the general trait of 'sharing', or egalitarianism, in all hunter-gatherer groups and particularly in the San of Africa remains sure. It was also widely observed in the forest-dwelling Pygmies, the

San's closest relatives. Sharing and the drive to harmony within these groups were survival imperatives and, while they lived in a style which had no concept of private property or fixed territory, they were retained and developed by intelligent Late Stone Age people over thousands of years. Without a 'sharing' culture, applied with some differences as geography dictated, they would not have survived.

*

Speculations about the Khoi, the branch who adopted a pastoral life with acquired domestic animals continue today and in 2006 Prof Felix Chami of the University of Dar-es-Salaam introduced a far-fetched idea in his book, *The Unity of Ancient African History 3000BC - 500AD*. He postulated that the Khoisan were brought to southern Africa by Hanno's Carthaginian colonising voyage in 410BC. The concept of the Khoi pastoralist or herding culture being introduced to the Cape from various exotic locations such as Persia or the Mediterranean was seriously considered by European colonial historiographers and it is interesting that it has been resuscitated by a native-African academic.

    Chami's thesis is prompted by his conclusion that the Khoisan as a whole have greater cultural affinity with the Berber/Libyan people of the Maghreb than to the Negro people of the central, tropical regions of Africa. The corollary of this is that the direct descendants of the ancient Core-People of Africa, those that remained behind after the out-of-Africa migration of 80,000 years ago, have to be the Negroid peoples and that the Khoisan are descended from mainly Afro-Asiatics who infiltrated back along the Mediterranean probably from about 50,000 years ago. This concept is contrary to the genetic evidence which shows that the San-Bushmen race inhabiting southern Africa, most particularly those who lived in the Kalahari, are the least hybridised on Earth and the evidence follows.

    I understand Chami's thinking because it coincides with my belief that there was a common hunter-gathering savannah culture throughout Africa at the time of the out-of-Africa migrations of 80,000 years ago which prevailed until the neolithic agricultural revolution and the birth of the northern hemisphere civilisation after the last ice age. Considerations of culture have to be separate from race, however, and if the Berber people on the northern side of the Sahara displayed vestiges of the universal African savannah culture two and half millennia ago, similar to that exhibited by the Khoisan at the other end of Africa in a similar time-frame, that does not in any way mean that they were transported there by Carthaginians, or in some lost and romantic migrating horde as 19th century writers proposed. Chami was also influenced by evidence that the out-of-Africa migrations of 80,000 years ago were undertaken by people of north-eastern Africa who had diverged genetically by then from the least hybridised African race, the Khoisan.

An important report by 21 scientists including Peter Underhill and L. Luca Cavalli-Sforza in *Nature Genetics* in 2000 summarises one genetic study at that time:

> Binary polymorphisms associated with the non-recombining region of the human Y chromosome (NRY) preserve the paternal genetic legacy of our species that has persisted to the present, permitting inference of human evolution, population affinity and demographic history. We used denaturing high-performance liquid chromatography to identify 160 of the 166 bi-allelic and 1 tri-allelic site that formed a parsimonious genealogy of 116 haplotypes, several of which display distinct population affinities based on the analysis of 1062 globally representative individuals. A minority of contemporary East Africans and Khoisan represent the descendants of the most ancestral patrilineages of anatomically modern humans that left Africa between 35,000 and 89,000 years ago.

Apart from work done by Phillip Tobias published in 1972, more recent studies confirmed a clear distinction between Negro and Khoisan. One in particular is by Bieber, Heidi; Bieber, Sebastian W; Rodewald, Alexander & Christiansen, Kerrin : *Genetic study of African populations: Polymorphisms of the plasma proteins TF, PI, FI3B, and AHSG in populations of Namibia and Mozambique* (Human Biology, Feb 1997). They state:

> The results show that indeed the systems TF, PI, and AHSG are of high value for anthropological genetics. The allele frequencies for these systems enable clear identification of and distinction between black African and

> Khoisan populations. The F13B locus, on the other hand, reveals for both the black African and the Khoisan populations specific and unique African variants: a high frequency of F13B*2 and the lowest frequency of F13B*3 so far worldwide. The new data are compared with results for TF and PI in another black African population of Mozambique, which Rodewald et al. (1988) had studied previously. The dendrogram, based on genetic distance data D and average linkage cluster analysis, shows minimal distance between both black African populations of Namibia and Mozambique and marked distance between those and the Khoisan population of Namibia.

These studies show that the out-of-Africa migrants were descended from a core people from whom modern East African Negroes and southern African Khoisan are also descended. Chami's speculation claims the Khoisan were descended from a retro-migration back along the Mediterranean from the Levant and therefore it follows that they would also have carried the original out-of-Africa genes. But, all modern people all over the world carry those ancient genes, and Chami's thesis (and any other grand idea of migration or invasion from the Mediterranean) collapses on the evidence that the Khoisan people separated genetically from the rest of Africans at a very early time; earlier than the out-of-Africa migration. The Khoisan is the most ancient genetically identifiable race today. Stephen Oppenheimer in his exhaustive examination of the out-of-Africa migration, *Out of Eden* (2003), examines the genetic evidence in detail.

In my book, *Seashore Man & African Eve*, I explored the problem of racial diversion in the period 160,000 to 35,000 years ago in the light of latest genetic studies, and I repeat here some of what I wrote.

According to recent genetic studies, the divergence of the Khoisan from the rest of African peoples had occurred at 150,000 BP with the appearance of the haplogroup L0 which descends from R0, while the rest of African peoples are of the haplogroup L1 with numerous subdivisions over time.

Doron M. Behar, and his several colleagues wrote in the Introduction to their important paper, *The Dawn of Human Matrilineal Diversity*, in *The American Journal of Human Genetics* 82, 1130–1140, May 2008:

Current genetic data support the hypothesis of a predominantly single origin for anatomically modern humans. The phylogeny of the maternally inherited mitochondrial DNA (mtDNA) has played a pivotal role in this model by anchoring our most recent maternal common ancestor to sub-Saharan Africa and suggesting a single dispersal wave out of that continent which populated the rest of the world much later. However, despite its importance as the cradle of humanity and the main location of anatomically modern humans for most of their existence, the initial Homo sapiens population dynamics and dispersal routes remain poorly understood. The human mtDNA phylogeny can be collapsed into two daughter branches, L0 and L10203040506 (L105), located on opposite sides of its root. The L105 branch is far more widespread and has given rise to almost every mtDNA lineage found today, with two clades on this branch, (L3)M and (L3)N, forming the bulk of world wide non-African genetic diversity and marking the out-of-Africa dispersal 50,000–65,000 years before present.

This entangled pattern of mtDNA variation gives an initial impression of lack of internal maternal genetic structure within the continent. Alternatively, it might indicate the elimination of such an early structure because of massive demographic shifts within the continent, most dominant of which was certainly the recent Bantu expansions [in the last 3-5,000 years] and spread of agriculturist style of living. However, some L(xM,N) clades do show significant phylogeographic structure in Africa, such as the localization of 1c1a to central Africa or the localization of L0d and L0k (previously L1d and L1k) to the Khoisan people, in which they account for over 60% of the contemporary mtDNA gene pool.

A most important fact emerges from this study. It is that the people who migrated out-of-Africa and populated the rest of the world were not of the clearly defined Khoisan race, with the fewest genetic divergences, carrying the L0 genetic marker. They belonged in the M and N clades of the group L3 which diverged from L2 at about 115,000 BP, which in turn diverged from L1 at about 165,000 BP which descended from the ancestral 'Mitochondrial Eve'. They were what may loosely be defined as ancestors of Negro or Negroid people.

All of these people were of the African Middle Stone Age which is observed in its clearest form in archaeological sites from the Cape, around East Africa and into the Sahara. The people of the savannahs who were least affected by the rigours of changing West and Central African forests kept to the least diluted line of cultural evolution. When the Sahara was re-populated during the wet time after 10,000 BP, culture flowed in from the south-east no matter who the people carrying it were genetically. The out-of-Africa migrants of 80,000 years ago may be genetically designated Negroes of the L3 haplogroup, but the culture they carried was that of the eastern African savannah from where they jumped off into Asia.

*   *   *

In very much more recent time, long after the out-of-Africa migrations, during the last 3,000 years, the herding branch of the Khoisan, Khoi, was absorbed or displaced by Bantu-speaking mixed agriculturalists everywhere except in the arid regions of western South Africa, southern Botswana and Namibia and along the winter-rainfall southern Cape coastal belt which Iron Age Bantu-speakers found to be unsuitable for settlement or colonisation.

Some historians and anthropologists propose that it was only in these regions that the Khoi lived and practised herding, and this is based on archaeology and the historical records of the first European colonists. What is not known with any certainty was whether Khoi inhabited lands which were later colonised by Bantu-speaking people. Archaeology, genetics and linguistics show that Bantu-speaking, Iron Age people moved in and absorbed or displaced hunter-gathering San. It is unclear whether any of the Khoisan in the lands which were later occupied by Bantu-speakers were herders, or Khoi, before absorption or displacement.

The conversion from hunting to Late Stone Age pastoral herding was a huge evolutionary jump on the way to the full range of agriculture. All the way down Africa, that evolutionary 'wind of change' gently surged, changing the people of the continent. But in most of Africa that fascinating intermediate phase of Stone Age pastoral society can be viewed only from archaeological sources and indirect extrapolation. Many South African and Namibian Khoi, however, were living in pastoral style well into the 19th century. Observers were still encountering isolated pockets of traditional Khoi herders in remote and arid parts of South Africa in the 1920s.

Monica Wilson, who tackled the Khoisan in some depth in the *Oxford History of South Africa* (1969), wrote:

> The picture is a more complex one than when it was supposed that language, economy, and physical type were inextricably linked. There were small groups of hunters scattered throughout the country. Some living on the sea-shore were particularly adapted to fishing; others were adapted to the desert; some spoke a San language, i.e. a language peculiar to hunters; others KhoiKhoi. Adaptation to the desert was not correlated with one type of language rather than another, and there was no reason to suppose that fishing was either. There were also groups of herders, most of whom spoke KhoiKhoi, but some [the Damara] ... spoke a Bantu language, Herero.

Monica Wilson compiled from earlier authorities the following main Khoi clan and language groups which were identified in European colonial times: Chochoqua, Chariguriqua, Chainoqua, Hessequa, Gouriqua, Attaqua, Outeniqua, Ingua, Damaqua, Gona, Griqua, Kora and Nama.

Apart from much which can be read in official reports, journals and court records, there were detailed contemporary private observations from the Cape colonial period, and here is a particular example, part of a letter written by Mr J. Maxwell to the Rev. Dr. Harris FRS in 1708:

> The Hottentots [Khoi], natives of this place, are a race of men distinct from both negroes and European whites, for their hair is woolly, short and frizzled, their noses flat and their lips thick, but their skin is naturally as white as ours, as appeared by a Hottentot child brought up by the Dutch in their fort here [kept out of the sun]. Their stature is universally of a middle size. They are clean limbed, well proportioned, and very nimble. I never saw a fat person among them.

> They besmear their faces and bodies all over with suet and other oleaginous stuff, which together with exposing their bodies to a warm sun, makes their skin of a tawny colour, and causes them to stink so that one may smell 'em at a considerable distance to windward. They adorn their hair, which is always clotted with grease and nastiness like the thrumms of a mop, with shells, pieces of copper etc. Both sexes are clad with skin commonly of a sheep, but sometimes such wild beasts as they happen to kill, the hairy side outward in summer but inward in winter, off which I have seen 'em pick and eat the lice in the streets. The women wear skins cut in thongs about their legs to the length of a great many yards, which when dry with the inside out look like so many sheep's guts that most strangers have mistaken them for such. The men hang their privities in a bag, and the women cover theirs with a flap or apron made of skin. The women wear a cap of skin first dried and stitched together whereas the men commonly go bareheaded. They go barefooted except that when they travel they wear a piece of skin fastened about their feet. Their weapons are javelins with which they are very dexterous at hitting the mark, and bows with poisoned arrows, which kill as I am informed upon drawing blood, but what they are envenomed with I could not learn. Their houses are hemispherical, made of mats supported by stakes, so low that a tall man cannot stand upright in one of them. These they remove upon occasions as the ancient nomads did their tents.

The problem for historiographers is that the Khoisan were the remnants of Late Stone Age aboriginal Africans occupying maybe half of the continent at one time, but who have had minuscule dynamic impact on the history of sub-Sahara Africa in the last 3,000 years. Since their presence (like the Pygmies) was passive and benign until the last three centuries and their direct influence on history seems confined to the cultural attributes of magnificent rock-art and the addition of words and 'click' sounds into the Bantu languages of the people who absorbed them, they lack excitement. They were not cannibals, did not wage aggressive wars, build stone towns, migrate in tribes nor colonise others.

As any story-teller knows, without conflict and violence there is no drama.

Because the Negroid Bantu-speaking migrations and colonising activity absorbed the Khoisan peoples as they moved down Africa, it has been difficult for academic writers to engage the problem of the acquisition of exotic domestic livestock and pottery culture by the Khoi without speculation. Earlier colonial historians such as George McCall Theal explored theories in great detail. Some of their romantic Victorian ideas, such as the possibility of a migration directly from the Middle-East or even Persia, were ridiculed in the light of new knowledge and therefore much of their sounder speculations may have been neglected. However, recent research and scholarship on eastern Africa shines more light.

Dr. Richard Wilding, an archaeologist who spent much of his career working in and studying north-east Africa, wrote in a substantial paper on East Africa, *The Shorefolk* (1987):

> There may have been Cushitic speakers [from Ethiopia] in the plains behind the coast [of Kenya] before the end of the Late Stone Age. Stone tools reminiscent of those from the Late Stone Age are still manufactured by Cushitic speakers in southern Ethiopia .... and southeastern Ethiopia .... These are used for hunting and gathering, including the preparation of hides. Such tool connections with southern Ethiopia may be discernible as early as the sixth millennium BC. ... There are some linguistic data which may encourage this notion ... that southern Cushitic speaking pastoralists moved onto northern rangelands from southern Ethiopia about four thousand years ago, intermingling with the existing hunting and gathering communities.

I met Richard in 1987 and we had a wide-ranging conversation about the movement of peoples in eastern Africa from say 3,000 to 1,000 years ago. I roughly recorded part of that conversation in my diary and reproduced it in my book, *Two Shores of the Ocean* (1992).

> "Are we sure that there were great migrations?" he said. "And were they exclusively Bantu-speaking? Many people have taken for granted that there was some kind of wave advancing down Africa two millennia ago like the Asian Barbarians swamping Europe. I don't see evidence for that. In the first place, there were too few to make waves

and Africa is not so hospitable that the geography welcomes tidal waves flowing over the landscape. I reckon it was thin trickles following the best paths; some succeeded, others died out. It was never coordinated. Small groups wandered across Africa, farmers kept to farming country or starved; nomadic pastoralists followed good grass where it existed, like the wild game. They must usually have coexisted and there was no reason to clash. Africa is very big. And if there was a clash it would have been a minor skirmish with lots of sound and fury but little violence."

"And if there were no clashes or struggles for territory they were friendly to each other and traded?" I prompted.

"They helped each other out," he replied. "Firstly because they were facing a vast and often dangerous environment and secondly because they had things for each other. Farmers needed animals for milk, ceremonial rituals and for barter or to buy brides. Nomadic pastoralists needed grains, metal and other artifacts. Metal-working clans attached themselves to one or another farming group where there were raw materials, in a symbiotic arrangement. Recent examples suggest that culture was generally similar and the principles were universal but each group followed the one formed by the environment of their origins.

"Pastoralists from Nubia had a rather different life-style to farmers from Lake Victoria and so on," Richard continued.

"Customary cultures were carried with them as they moved. As time went by there were amendments and changes where necessary and, where it was not necessary, they stuck to their cultural heritage. There is some evidence for the transference of elements of Sudanic languages to Africa south of the Congo. So, probably the movement was not exclusively Bantu-speaking either. We must remember that there is evidence of Cushitic-speaking pastoralists from Ethiopia in East Africa before the Bantu-speakers came. And those Cushitic-speaking pastoralists were probably also occasional Neolithic agriculturalists where it suited, as they have been recently. We are sure that Cushitic-speakers were in Tanzania, but who knows how far south they trickled either before the movements of Bantu-speakers or together with them? Or later?"

From Richard Wilding's *The Shorefolk* (1987):

> Cushitic speaking groups [from north-east Africa] seem to have been as far south as the hinterland of the Tanzanian coast by the first millennium AD .... The spread of Cushitic speakers over the rangelands behind the northern East African coast seems to have been fairly general. The mechanisms for the spread are not known, but modern ethnographic study might suggest a gradual diffusion of related peoples, pressed by bad spells of drought or disease, and occasionally propelled short distances by war, trickling in small groups into neighbouring communities, or slipping through the interstices of hunting and gathering communities, intermarrying the while.
>
> The movement appears to have originated in south eastern Ethiopia for reasons not yet entirely clear .... The linguistic evidence suggests that there were imbrications of east Cushitic speakers over south Cushitic speakers during the first millennium.... This has usually to mean that there were new people moving into the region. It may mean simply a linguistic evolution within a stationary and stable population. ... It would seem that the notion of small immigrant groups is the more likely explanation....
>
> ... The whole linguistic exercise presents serious historical problems of dating until the performance of glottochronological techniques appears more reliable: nevertheless the sequence of population overlays in the region [East Africa] is most interesting, and the period of a thousand years or more after the time of Christ seems very plausible.

Professor Roland Oliver in *The African Experience* (1991) gives his summary of events that led to the evolution of a Khoi society:

> It may even be that, before the Bantu closed in to the south [of East Africa] there was some direct contact between the southern Cushites and the most northerly of the ancient Khoisan peoples, through which the earliest domestic animals were able to reach southern Africa in the hands of

people who were not yet cultivators. More likely, however, they passed through the hands of early Bantu intermediaries, reaching the Khoi of northern Botswana and western Zimbabwe by about the first century BC.

Supporting the second possibility, Oliver pointed out that remains of cattle and sheep have been found at excavations in the savannah belt to the south of the Congo rainforest with dates from the third to first centuries BC. Cattle and sheep seem to have been introduced from East African high grasslands to the lands of the upper Zambezi tributaries in Zambia several centuries before evidence of the movement of Bantu-speaking migrants further to the east. The people who owned these cattle and sheep, and who may have been one of several sources of livestock for the Khoi and the transference of herding culture, were Late Stone Age cultivators who kept livestock. Perhaps they were Khoi who had migrated there with livestock after acquiring them from Nilotic or Cushitic people, or both at different times, further north. Perhaps there was a movement of both ancestral Khoi and Nilotic or Cushitic people southwards in advance of any Bantu-speaking migrants? New pottery evidence from East Africa, provided by Prof. Felix Chami's archaeology since 2000, definitely proves an earlier infusion of agriculturalists than had been accepted years ago. Richard Wilding's speculations, aired in conversation with me in 1987, were substantiated by Chami's archaeology.

At the wide-ranging and unique conference, *The Growth of Farming Communities in Africa from the Equator Southwards*, at Cambridge in 1994 there was little attention given to Late Stone Age agriculture and none to the Khoi. Presumably, there was insufficient fresh archaeology or other data then on which to base papers. But lately I have been conscious of an increased interest in this problem of the first transformation of the Late Stone Age to agriculture in southern Africa, stimulated by extended studies in South Africa and the particular work of Felix Chami in Tanzania. Chami provides definitive evidence of pre-Bantu, Late Stone Age agricultural communities on the Tanzanian coast from pottery at sites dated to the second and third centuries BC. He identified cattle bones in the Kuumbi Cave on Zanzibar in the Late Stone Age layer.

Dr Karim Sadr of the University of Botswana prepared a paper which is well regarded and published in the *African*

Archaeological Review in 1998. He summarised the archaeological evidence from pottery and animal fossils and the intellectual arguments regarding the arrival of herding culture and the Khoi at the far end of Africa. In the conclusions to his paper, *The First Herders at the Cape of Good Hope*, he refers to the two alternative ways in which Late Stone Age herding culture reached southern Africa. They are by migration of herding people and diffusion of knowledge and practice through existing hunter-gatherer communities. As discussed earlier, the conflicting possibilities of migration and diffusion are a perennial issue when considering the movement of culture in Africa. Sadr wrote:

> The textbook Khoi migration hypothesis would have been supported if, first, the earliest livestock and pottery had consistently appeared together as a package in LSA [Late Stone Age] sites of southern Africa and, second, a stylistic chain linked the earliest pottery from the Cape to northern Botswana. The clear regional diversity in the earliest ceramic styles and the unsynchronized appearance of livestock and pottery instead favour the alternative diffusion hypothesis. The diffusion scenario has received added support from evidence which suggests that the Khoi may have arrived at the Cape around the end of the first millennium AD, long after the first sheep and pottery reached the southern tip of Africa.

However, within the paper Sadr describes in detail the objections to a purely diffusionist theory on the grounds that hunter-gatherers, because of their powerful inheritance of a 'sharing' egalitarian lifestyle, would find it difficult to adopt a herding culture which implies personal property. He concludes:

> Diffusion is thus considered theoretically impossible: livestock and pottery must have reached the Cape by a migration of herders.

Thus there was a lack of clear consensus and unless there are more archaeological finds, with the necessary wide spanning of time and place, it may be impossible to find scientific proof of how and when herding and the beginning of agriculture first spread on the savannahs of central-southern Africa But there is so much of the pre-history of mankind which is unproven that this bothers me not at all. Indeed, if all the speculations about the evolution of mankind had been settled, there would not be so much fun in pursuing the answers!

The paucity of fossil and pottery evidence available today tempts an over-simple historiography of the Khoi and their ancestry, and the source of herding culture in parallel to hunter-gatherers of the same race and genetic heritage in this huge region of Africa, before the arrival of more organised cattle-herding with mixed agriculture of the Iron Age carried by Negro Bantu-speakers. But we are looking into at least two thousand years and possibly as much as three thousand years.

For a start, I cannot possibly accept that diffusion is unlikely or impossible as a method for the movement of herding culture on the grounds that hunter-gatherers would have difficulty in changing from a sharing ethos to a limited property-owning ethos. If this was not possible in several hundred years in southern Africa, how did it ever occur anywhere?

Earlier, I have suggested that hunting may have translated to herding more easily than gathering to cultivating because of intimate knowledge of herd animals as prey since the mists of time. A change in technique towards animals was rather more easy than a radical change in society from sharing to property-owning.

Simply, it could have been a gradual transition with the first experimental herds being held by the community in common, then entrusted to the most expert family while others learned and numbers of animals expanded until a natural adoption of separate family-owned flocks or herds within the group evolved.

Hunting was a communal male activity, but not all men in a band were equal in their prowess. There were always the better hunters and the stumblers. Gathering by women and children was a shared activity in principle but the spoils were not shared unless there were unneeded surpluses or necessity required it. This is naturally a common practice amongst families or groups of friends today on any holiday picnic or camping trip.

The rest of the numerous communal aspects of life continued with little change and there was a prevailing culture of sharing no matter whether this was applied in detail to particular activities. No doubt there were tensions and conflict during transition, but 'survival of the fittest' feedback eventually resolves these situations in the wilderness with surprising rapidity. The diffusion of the technique of herding would have been accompanied by diffusion of advice and understanding of the social consequences. Khoisan were especially notable for their endless discussion and internal communication on every issue of life, which was an integral part of their communal lifestyle. They lived close together, intimately, all the time.

Anthropologists have noted in other contexts, and Sadr covers them in his paper, that the San hunter-gatherers exhibited considerable flexibility when confronting different problems, as I would expect from consideration of the enormous power of their continual intimate daily communion. Every aspect and problem of living was thrashed out in finest detail sitting around the camp fire before decisions were taken. They had the enormous unifying strength of their singing and dancing and guidance from the trance-induced visions of their spiritual leaders. Every observer in historical time wrote about their fondness for frequent singing and dancing together. It was recorded how they had an easy system of coalescing into 'hordes' under a trusted chieftain when convenient for the movement of flocks in season and splitting up again when it was no longer necessary for mutual help or protection.

These Khoisan people, both hunter-gatherer San and herding Khoi, in historical time showed that they were old masters of their environments and the social cultures needed for that mastery. Judging them by our criteria for formal 'civilised' society, bound by laws and rigid structures, is not sensible. This ancient ability to change in the face of environmental or social challenge by mutual consensual agreement within all members of a family, clan or tribal group is a particular African trait displayed by all the different cultures across the major racial divisions. When the group became too large for all members to take part, it was a conclave of elders which gathered for days of talk. Demagoguery can be present in any community when a charismatic leader emerges, but ruthless despotism by tribal chieftains in Africa such as seen with King Shaka in Zululand in the 19th century were exceptions or aberrations. Chiefs ruled by consensus of the people and if they were seen to have failed their people they were deposed and sometimes ritually put to death. This universal tradition has been commented on by both medieval Arab geographers and modern Europeans.

European colonial administrators jibbed at the apparent lethargy or indeciveness of Africans when asked to consider new ideas. They had to have sufficient time to consider properly. It was the ancient strength of African society brought down by unbroken lines of tradition. These traditions were not damaged as they were by the great traumatic changes into urban national structures which were endured in the Indus, Mesopotamian, Middle Eastern and Nile civilisations.

The mechanism of clientship must also not be forgotten. Diffusion of ideas and knowledge did not progress only by observation or propinquity. It probably moved fastest by one group having to become clients to another in the face of disaster or trial. I believe that clientship was a force in Late Stone Age societies and changes within them that is often given insufficient attention. Hunter-gatherers forced by natural circumstances to become clients to herders learned thoroughly.

By the sixteenth century, the larger territorial regions divided amongst Late Stone Age Khoisan and Iron Age Negro people had long been stabilised. Each group occupied the lands best suited to their economy; the Bantu-speaking Negroes needed adequate summer-rainfall zones for their main food-crop, the sorghums and millet cereals, and for the healthy grasslands needed by their cattle. Tsetse-infested bush was no good to them, about 600mm of rain per annum was the minimum and there had to be perennial water. A certain amount of local nomadism, the moving of herds to winter grazing, was acceptable.

The Khoi did not have the same need of rainfall, but they had to have perennial water and seasonal grazing. Therefore they could tolerate the arid conditions which were of no use to the Bantu-speakers, but had to move almost constantly in order to conserve the vegetation of their ranges. They tended to keep to the valleys of major rivers and the lands they found suited best were around the southern end of the Cape where there was a Mediterranean-type climate of winter rain.

The San could survive almost everywhere but kept away from the belligerent Bantu-speakers and Khoi who were protective of their jealously-guarded herds. The San had found it advisable to live where domesticated flocks and herds could not easily survive: in the deserts and mountains.

It is difficult therefore to plot precisely the first movements of herding culture from the north down to the south. It cannot be assumed that the Khoi first met in 1488 by the Portuguese were the particular people who were responsible for this cultural migration. It is entirely likely that they were the inheritors of a second or third, or even fourth, 'wave' of culture originating somewhere in central or eastern Africa. And the movements of those cultures are subject to the controversy of migration versus diffusion, or a combination of both either sequentially or in parallel, or both!

Dr Karim Sadr covered the archaeological and linguistic issues in his paper. The principle problem is that fossil evidence of exotic sheep have been found with dates of about 2,000 years ago in distinct sites on the southern shores of the Cape but pottery does not necessarily align itself with fossil bones. There was a more complex movement of people around in southern Africa in the last 2,000 years then a simple migration of herders. Sadr:

> There is little doubt that small livestock and pottery reached the Cape at least 2,000 years ago. Sheep bones from the Late Stone Age (LSA) sites of Spoegrivier and Blombos, respectively, in the northern and southern Cape have been AMS dated to ca. 2105, 1960, and 1880 B.P. [before present], while several sites such as Hawston, Die Kelders, Boomplaas and Kasteelberg A have yielded pottery in layers dated to the first few centuries AD.

That seems satisfactory for a migration theory, but it is not simple. Again, Sadr:

> However, AMS dating shows that the sheep bones at KBA (and Die Kelders) are younger than the radiocarbon-dated pottery-bearing layers. Conversely, at Blombos on the southern Cape coast, sheep bones were found in a layer underneath the first pottery, but the AMS date of the bone overlaps with the radiocarbon date of the pottery level.
> ... Farther north in Namibia, the Falls Rock Shelter and Snake Rock sequences contain pottery long before the appearance of sheep. ...
> .... Given that this is the current state of knowledge, it can be stated that so far the evidence provides a better fit to the proposition that sheep and pottery diffused at variable rates. Had the Khoi migrants

brought sheep and pottery, then both traits ought to have appeared together regularly.

Argument about diffusion versus migration may be interesting for academic precision, but I favour a mix or matrix of processes anyway when one is considering long periods of time like 2,000 years. What is clear, at least to Karim Sadr, is that there were at least two identifiable activities, possibly attributable to at least two separate groupings of Khoi people. There were herders with pottery and herders without.

The consideration of pottery from the Late Stone Age and its relationship to the distinctive styles of the Early Iron Age is a complex matter. As I have said elsewhere, it is cultivators who had to have pottery to store and cook their grains and tubers; herders do not need pottery because leather or skin bags have been used by any number of people all over the world for carrying and storing liquids. But, pots are indeed better utensils for storing and carrying milk or its processed derivatives than skin bags. Pottery is a sophisticated luxury to herders because it is a nuisance if they are on the move, for pots have to be carried or discarded and new ones made or traded at the next stop. Pottery indicates pack-animals and sporadic settlement for a reasonable time.

The difference between Late Stone Age and Early Iron Age pottery was pointed out to me by archaeologists John Kinehan and Leon Jacobson when I met them for discussions in the 1970s. Dr. Kinehan excavated the important Falls Rock Shelter site in Namibia. Dr. Leon Jacobson, an archaeologist with years of fieldwork in southern Africa told me that the pottery which pastoralists made in southern Namibia was of finer quality than that associated with the Late Iron Age of 1200-1500 AD. He said: "It is beautiful, thin-walled, well-fired, burnished pottery with distinctive patterning; pointy-bottomed, some with lugs. It was not used for cooking, but for storage." His face was lit with enthusiasm as he spoke. "At the same time, ostrich shell jewellery was small and fine in comparison to the coarser work of the Iron-age." Sadr confirmed the differences with more and recent examples.

> All these wares [Late Stone Age] are thin walled and well fired. Vessels are quite small. Spouts, a common feature of the earliest ceramic assemblages at the Cape, are occasionally found in early Namibian and

Botswana/Zimbabwean assemblages. Indeed, they can be found as far north as Kenya and Nubia.

I find the pottery vessels most illuminating. Not only do spouts immediately suggest the use of the pots for liquids, significant for herders, but here is the hint of a general herding culture connection along the whole length of eastern-southern Africa at 2,000 years ago and before. Lugs, which apparently appeared maybe a thousand years later, clearly indicate that they were slung from pack-animals and that this was a common enough activity to make vessels with those features as a standard. It seems clear to me that vessels with lugs for slinging them on pack-animals shows when cattle became common. This particular change in pottery style could thus be of considerable significance.

I assume from this evidence that the first people who introduced herding culture to the far south of Africa had sheep, occasionally with pottery, and probably practised what Sadr calls a 'hunter-herder economy'. That is, they had a kind of transitional economy where most food procurement was through traditional methods and sheep or goats were communal assets. A second cultural change was more definite. Cattle were prominent and used as pack and riding animals, the culture included a larger range of tools and equipment, requiring the need for pack-animals when on the move, and the social system was probably that of conventional pastoralists with family groups owning flocks and herds which join with other families into clan groups, ultimately into some occasional larger loose federation. In Iron Age agricultural society this structure evolved into a formal and disciplined state with a hierarchy of elders and chiefs but was never adopted by the Khoi.

Decoration is important in divining the relationship between people and their ancestry. Sadr has this to say:

> Clearly, the ceramic decorations from Botswana, Namibia, and the southwestern Cape are quite different. There are also differences in other stylistic aspects of vessels. ...
>
> In addition, overall vessel shapes from Botswana and the Cape differ radically ...
>
> ... It is clear that there were pronounced regional stylistic differences, although a few sherds do suggest interregional contact: ...

There is no quick and simple solution to be seen of one group of people migrating down Africa with sheep. Whether by diffusion though the San hunter-gathering communities or by particular movements of established herding clans, those animals and the people who herded them were at the end of Africa by 2,000 years ago. What is also clear is that there were different groups of people moving about in southern Africa from 2,000 years ago who had pottery but had different cultural allegiances.

I am sure in my mind that there were a number of different movements of people at different times and the sheep and the techniques of husbandry and associated pottery industry were both carried by migration and diffused to some hunter-gatherers either by propinquity or by clientship.

This may seem to be a very superficial overview encompassing all possibilities, but it is dictated by the limited facts as we know them now. Karim Sadr analyses the meticulous work done on the archaeological sites and theoretical discussion carried out by himself and other archaeologists and anthropologists in the last twenty years or more in his authoritative paper. At least the romantic myth of a great surge of migration by some strange Hamitic or other colonising race has been laid to rest forever.

*   *

The matter of the Khoi in historic time, the last 500 years, is clearer. Early colonial records and descriptions describe the several clans with language differences who lived along the well-watered southern Cape mountainous regions (quoted above), but which were not occupied by Iron Age Bantu-speakers because of their unsuitability.

Linguistic research, oral history and historical record provided a reasonable exposition of the migration routes of the Khoi who were found in particular regions by European settlers and explorers. Through the changing seasons, which are very clear-cut in the Cape from wet cool winter to hot dry summer, the Khoi moved their flocks and herds but each group of linguistically related clans, which belonged to a loose federal union sometimes described as a 'horde', kept to a roughly defined territory.

One of the Khoi's own stories of origins is that they came from the east, from the sun, where there was much water. Until archaeology provided greater detail, especially regarding the great extent in time during which herding was prevalent, there was a generally accepted story of origins. That simple narrative was that the Khoi came from a misty heartland, or heartlands, north of the Zambezi river.

The generally-accepted historiography of the 1960s and 70s continues. With cattle and sheep they moved into southern Africa, generally following river roads : the Zambezi-Chobe-Okavango and the Limpopo-Vaal-Orange. They did not follow the Zambezi or Limpopo into the eastern lowveld for the obvious reason that it was a tsetse-fly zone anathema to cattle. Having been carried by geography to the Atlantic side of the sub-continent, some settled there. Others migrated in the short wet seasons down the west coast to the southwest corner and then along the southern mountain zone until they met Negro Bantu-speaking Iron Age agriculturalists with their cattle-cult and came to a stop. Movement from the river roads which they had followed into the healthy highlands to the southeast, following other rivers, was impossible because those lands were occupied by other Bantu-speaking tribes.

Prof. Monica Wilson in the *Oxford History of South Africa* (1969) summarises the Khoi from observations in the 17th and 18th centuries.

> The Khoikhoi herders had large flocks of fat-tailed sheep and herds of cattle, and milk was their staple food, men drinking cows' milk only, and women and children that of ewes. The Nama alone, who were trading with the 'goat people' - the Sotho-speaking Thlaping - had goats. The cattle were the long-horned type, ancestral to the modern Afrikander strain, and they were numerous in proportion to the men. Van Riebeeck [17th century] speaks of a camp of 'Saldanha Man' with fifteen huts and a population of about two-hundred-and-fifty, men women, and children, with fifteen or sixteen hundred cattle and sheep besides - six head of cattle per person. Another horde had eleven to twelve hundred cattle and six hundred sheep. ...
> ... The herders did not ordinarily kill stock for meat, but only on the celebration of rituals, and they depended for food not only on their herds but on hunting, fishing and collecting of veldkos [literally : field food] and honey. When meat was plentiful they dried it to make biltong, and honey was used to brew mead. ... Early

travellers described both men and women carrying loads on the back. This is what forest people do, further north, and perhaps it is characteristic of hunters and collectors, whereas among cultivators, living in open country like the Nguni and Sotho, the women carry loads on their heads.

They wore sheepskin for clothing and enjoyed the use of jewellery and decoration. Copper was mined and smelted by Khoi clans who lived close to the source in Namaqualand or Namibia in historical time. They prized beads used by Europeans for barter and did not care that much for iron utensils like knives. Oxen were used as pack-animals and for riding and bulls were used in warfare, herded and sent infuriated against an enemy.

Monica Wilson, with the knowledge available in the 1960s, wrote:

> Riding was common in the Sahara and East Africa, and Sofala [central Mozambique] in the tenth century. The absence of these techniques among the Nguni and Sotho peoples [Bantu-speaking language groups in South Africa] suggests that some Khoikhoi ancestors may indeed have learnt from these men of Sofala, and remained isolated from them thereafter, until the eighteenth century.

This is an interesting thought and follows the trail of evidence suggesting an early connection between Khoi, their cattle and East Africa. That Bantu-speaking people of South Africa did not ride cattle and that the Khoi did suggests a different source of cattle by these two people of different racial and cultural origins. The cattle themselves were apparently of different breeds and this is a matter I go into in the next chapter.

The division between Late Stone Age pottery with the arrival of lugs occurred at about 1,200 - 900 years ago, coinciding with a warm climatic period. This warm period also has importance when considering the social changes within Bantu-speaking society, with strong population movements, in most of eastern-southern Africa dividing the Early and Late Iron Ages. It would seem that the Khoi whom Europeans met from the late 15th century onwards had been subject to some major cultural shift. The centuries around 1,000 AD was a time of a general shifting about of all people in Africa. The particular manifestation of this movement within the Khoi is their pottery with lugs. Karim Sadr wrote:

> As a further test implication of the late Khoi arrival hypothesis, it is proposed that the appearance of the pierced lugs in the archaeological sequence of Namibia and the Cape should correlate with a sudden shift in other aspects of material culture. If the appearance of lugs is the only noticeable change, it would suggest diffusion of an isolated trait which was then accommodated into the existing, indigenous material culture. If, on the other hand, the lugs were brought by foreign settlers, there should be a concomitant disjunction in many other aspects of the archaeological record. Most archaeological sequences do not cover both the pre- and postlug periods. Only KBB in the southwestern Cape clearly correlates with a major change in almost all aspects of material culture. Radiocarbon dates bracket the cultural disjunction between ca. 1200 and 900 B.P.

Because it has not been possible yet at this or other sites to determine how fast changes occurred (slow change over a hundred years or more could indicate internal evolution), Sadr and his colleagues are reluctant to declare certainty. More evidence from more sites is necessary for proof. I am sure in my mind that there were new movements of people involved, but whether this was a wholesale movement bringing different styles with it, or the infusion of small numbers with those ideas which gradually spread cannot be decided.

*   *

Khoi were first met by Europeans in 1488 at Mossel Bay by Portuguese maritime explorers commanded by Bartolomeo Dias, the first Europeans to penetrate the Indian Ocean from the Atlantic around the Cape of Good Hope. Vasco da Gama's expedition to India nearly ten years later met San at St.Helena Bay, on the Atlantic side of the Cape, and Khoi at Mossel Bay. Thereafter, many sailors had contact with both San and Khoi along the Cape coast, and when the Dutch set up their colony at Cape Town in 1652 they relied on the Khoi to supply them with cattle and sheep by barter, a trade that continued for a century or more.

Brief descriptions of these people at St. Helena and Mossel Bay in late 1497 are the earliest first-hand accounts still available. From the *Diário da Viagem* of Vasco da Gama's flotilla (translated by Eric Axelson):

> In this land [St. Helena Bay on the Atlantic coast] the men [San] are swarthy. They eat only sea-wolves [seals] and the flesh of gazelles and the roots of plants. They wear sheaths on their genitals. Their arms are staffs of wild olive trees tipped with fire-treated horns. They have many dogs like those of Portugal and they bark the same as they do.
>
> ... He was small of body and looked like Sancho Mexia. He was going about gathering honey in the barren wastes, for the bees of that land place it at the foot of thickets. We took him to the Commander-in-Chief's [Vasco da Gama's] ship who placed him at his table, and he ate of everything that we ate. ....
>
> ... Friday [1st December 1497], while we were still in the said bay of São Bras [Mossel Bay on the Indian Ocean], there arrived about ninety men [Khoi], swarthy of appearance like those of St. Helena Bay; and some of them moved about on the beach, and others of them remained on the hills. ... On Saturday [2nd December] about two hundred negroes, large and small among them, arrived, bringing with them about twelve head of cattle made up of oxen and cows and four or five sheep; and when we saw them we went ashore at once. They at once began to play on four or five flutes, and some of them played high and others played low, harmonising together very well for negroes in whom music is not to be expected; and they danced like negroes. The Commander-in-Chief ordered the trumpets to be played and we in the boats danced, and so too did the Commander-in-Chief when he rejoined us. ...
>
> On Sunday as many men as before arrived, and they brought with them their wives and small children, who remained on top of a hillock close to the sea, and many oxen and cows; and they put themselves in two places alongside the sea and played and danced as on Saturday.

The Dutch colonists at the Cape in the 17th century recognised that there were two kinds of indigenous people of the same race, whom they determined later were clearly different to the Bantu-speaking Negroes who lived to the east and north.

For a hundred years, the San remained something of a curiosity for there was little contact. It was the relatively numerous Khoi with whom the Dutch and French settlers had most contact, firstly on the initiative of the settlers and the Cape government in order to barter cattle and sheep, and then later when they began sharing the same pasturage and territorial conflict arose.

In 1713 and 1755, the Khoi were decimated by smallpox epidemics that swept the land. The structure of Khoi culture was irrevocably damaged in the 18th century, but they recovered their numbers. Through miscegenation with mixed-race slaves, Europeans and Bantu-speakers, their genetic heritage was forever diluted, but in the 20th century the further one travelled from the old colonial centres of Cape Town, the western Cape and the Port Elizabeth metropolitan area, the closer one came to meet people who were little different in appearance to those who first met Europeans five hundred years before at Mossel Bay. Clientship to European settlers and the activities of Christian missionaries effectively degraded Khoi culture in the 18th and 19th centuries, but in recent years intense historical research has rekindled our knowledge. A useful summary of historical and archaeological information and its use in composing a clear picture of Khoi lives can be found in *The Cape Herders*, a handy book by Cape Town researchers, Boonzaier, Barens, Malherbe and Smith (1996).

Leaving aside the purity of ancient genetic and cultural descent, which was unrivalled in the San, until the middle 19th century some remote clan-groups of Khoi were unchanged living representatives of the Earth-shattering jump from hunter-gathering to agriculture which began in the Middle-East and the green Sahara. The sheep and cattle they herded were similar to those in the Sahara of 5,000 years ago and through them we might have gained greater insights about events at that revolutionary time.

They had no metallurgy apart from the easy production of copper from abundant ores, but had pottery and were at the peak of classical Late Stone Age industry. They did not cultivate which was the reason for their survival: they were able to exploit pastoral conditions to their limit, husbanding their precious livestock in marginal lands.

Many contemporary populist historians have given these aboriginal Africans scant attention in general historiography. Basil Davidson, for example, in *Africa, History of a Continent* (1972) has only this to say:

> Soon they [Dutch settlers] had turned the local Africans, who were Khoisans (Hottentots), into their slaves.

In his later book, *The Story of Africa* (1984), he is similarly brief and vague, though marginally more accurate though his naming is wrong:

> Two Khoi peoples - known to the new settlers (Dutch) as "Bushmen" and "Hottentots" - had lived in South Africa since Stone Age times, thousands of years earlier, ...
>
> The two Khoi peoples had yet to enter a "metal age", and were content to live as hunters and food gatherers, or, now and then, as masters of herds of cattle. Their way of life, organised in small family groups without any military means or potential, could present no obstacle to Dutch and then British conquest; and those living nearest to the Cape were soon reduced to slavery by the early settlers.

Davidson failed to make clear that though San and Khoi were of the same race and ancestry, they were economically, culturally and politically quite different.

Khoi when forced by external territorial pressures were capable of fighting fiercely, but they were not able to prevail against disciplined Bantu regiments or European firearms. There is historical record of warfare with other Khoisan, Dutch colonists and Bantu-speaking clans. As early as 1658 - 60 in the immediate vicinity of Cape Town there were skirmishes with Dutch settlers over pastureland and the Dutch settlement had to be fortified with a stone castle and a thorn fence with blockhouses around their territory as a defence against the Khoi.

At random, one can read of many battles and tribal skirmishes. There was a notable battle in 1775 between Koranna Khoi and San at the ford of *Kokounop* on the Vaal River near the colonial village of Douglas which the latter called "the place where no mercy was shown". Khoi Griqua clans, under the leadership of mixed-race runaway and freed slaves from the western Cape, became a major bandit force, raiding for cattle over a large area of the north-east Cape and Orange Free State in the 1820s and 30s. Khoi Nama bands terrorised parts of Namibia in the 18th century and plunged Herero-speaking cattle-herding clans back into a hunter-gathering life as refugees. Nama chiefs behaved similarly to European medieval barons in Namibia for a large part of the 19th century, being wily in diplomacy and when necessary going to war. Skirmishes and fully mobilised warfare erupted with their Herero neighbours from time to time. There was nothing simple, unsophisticated or passive about the Khoi of southern Africa. Koranna 'Hottentots' during the 1880s under their chief, Massouw, battled with both white Afrikaners and Bantu-speaking Sotho neighbours in the western Transvaal and hired a Boer commando under Sarel Cilliers as mercenaries.

The British raised Hottentot military and paramilitary formations on the eastern borders of the Cape Colony. Peaceable people became belligerent when territory became an issue under pressure from European colonists and general turmoil amongst the Nguni people of the eastern coastlands.

Davidson's extraordinarily over-simplified statements that the Khoisan were enslaved are quite incorrect. The only slaves held by European settlers in South Africa were procured by government licence and imported from elsewhere by the Dutch East India Company. It is one of the most important of South Africa's historical inheritances that Europeans never operated the slave-trade in South Africa. Anti-European commentators who exploit the trans-Atlantic slave-trade *ad nauseam* would be pleased to pin this evil on European colonists in South Africa but are unable to do so. M.F.Katzen in the *Oxford History of South Africa* (1969):

> The first slaves in any numbers were two shipments from Angola and West Africa [Guinea] in 1658 and 1659, but they were intractable, many absconded, and eventually nearly all were returned to the Company [Dutch East India Company]. ... in 1672 the burghers [settlers] owned only 53 slaves.
>
> ... the Company allowed importations to continue. Most Cape slaves were Africans, obtained

> mainly from Madagascar and East African slaving centres, including Delagoa Bay, where the Company maintained a slave depot from 1720 to 1730. Some slaves also came from the Bay of Bengal, Indonesia and other Asian areas until 1767. ... Although slaves were used throughout the Cape, they were concentrated in Cape Town and the western Cape, where they were domestic servants, skilled tradesmen and petty retailers (whose masters often hired them out, or allowed them to earn their own living on payment of certain sums), and farm labourers or herdsmen.

The landmark Cape Government Ordinance No. 50 of 17 July 1828, and its predecessors in 1787, 1803, 1809, 1812, 1819 and 1823, confirmed the status of 'Hottentots' as a free people and defined their various rights and obligations, particularly labour contracts with them, payment of wages and other conditions of employment.

Another pejorative generalisation that is common is that the Dutch and French settlers at the Cape tried to extinguish the Khoisan by deliberate genocide. Professor Jared Diamond, for example, in *The Rise and Fall of the Third Chimpanzee* (1991) asserts that the Khoisan were subject to the same destruction as the Aborigines of Australia and the plains 'Indians' of the North American interior.

Dutch and French settlers and Khoi fought for territory and both sides killed each other in raids on occasion, but there was no genocide by either side. The Khoi were too valuable as clients and servants, livestock traders and general middle-men for the settlers to wish to eliminate them even if they felt they had the moral right. Many white settlers had both informal and formal marriage relationships with Khoi women. And the settlers were too valuable to the Khoi as the source of manufactured goods and of employment. It was a complex inter-relationship which developed over two hundred years. Dr. Paul Maylam in *A History of the African Peoples of South Africa* (1986) remarked in regard to the colonial eastern frontier in the 17th century:

> Xhosa-grown dagga [marijuana] would flow westwards and Dutch-imported metal and beads would move eastwards from Table Bay, the Khoi playing a key intermediary role in this trading network.

The San, on the other hand, were hunted by Boer commandos after they raided settlers' cattle. But this behaviour was usually retaliatory, sporadic and clearly outside the law. Perpetrators were sometimes prosecuted although attempts by *landdrost* magistrates to get independent evidence from a frontier community was usually impossible. In the 1770s Afrikaner pastoralist expansion had been halted by San-Bushmen and it has been often generally stated by ardent critics of Afrikaner behaviour that this kind of situation was resolved by the extermination of the Bushmen.

But, archaeology has shown that on the upper Seacow River of Cape Province, as an example, this certainly did not happen and that Bushman bands and white Afrikaner settlers co-existed for at least another hundred years until their gradual absorption into the local 'Coloured' community. By contrast, in Australia land was given out when the Aboriginal occupiers were rounded up for re-settlement in camps or 'reserves'. Neither San nor Khoi were ever herded into 'reserves' in the South African colonies as the aboriginal peoples were in North America and Australia.

The conflict between Khoisan and Negro Bantu-speaking structures was more severe than it was with the few and far-scattered Boers. In earlier centuries, the Khoisan 'disappeared' in lands colonised by Bantu-speaking tribes. Their only observable surviving attributes are the 'click' sounds in the Bantu Nguni languages and their genetic trails. Famously, Nelson Mandela has Khoisan-origin genes. In historical time, as Maylam states:

> Many Khoi attached themselves to the Xhosa as clients, offering labour or military service in return for protection and security. Others resisted any form of Xhosa domination, preferring to enter the service of white colonists.
>
> Similarly diverse forms of interaction, violent and peaceful, occurred between the Xhosa and the San. There is evidence of great Xhosa brutality towards the San. But, as Peires has observed, there was also long-standing contact.

The proof of the relatively benign treatment of Khoisan people in South Africa by European settlers is the virile and burgeoning population of their descendants. Perhaps the clearest evidence of the differing fates of the Khoisan at the hands of European and Bantu-speaking colonists of South Africa is the descendants of these people. In areas settled by Bantu-speaking mixed agriculturalists between 250 AD and 1500 AD, there were no surviving Khoi. They were absorbed, or pushed aside. They survived in increasing numbers to the present day only in the Cape and Namibia where Europeans settled and where they came under the protection of European colonial law, however poorly it was often administered. Whereas in Australia the Aborigines were the only native people in the whole of the land, the Khoisan of southern Africa were already a dwindling minority under Bantu-speaking agriculturalists' pressures. The survival of Khoisan descendants in the large numbers they are today is arguably solely due to their protection and nurturing by European colonists.

The 'Cape Coloured' people, the descendants of the Khoisan and the products of mixing between them, the European settlers and slaves from East Africa and the Far East over a period of three hundred years, are a significant proportion of the present South African population, numbering about 9% or 4,000,000 in 2001. In contrast, the number of people claiming Aboriginal descent in Australia, many the product of mixed liaisons, was 1% or 190,000 in 2000.

There are many surviving firsthand formal and informal documentary descriptions of Khoisan and their cultures from the South African colonial period. It is for this reason that their strange neglect by historians and anthropologists in general surveys of Africa is most regrettable. Perhaps it is difficult for writers in the current political climate to admit that their existence is a testament to their survival only where they lived amongst European settlers.

*   *

In 1977 I was driving along a rough gravel road from Steinkopf in Namaqualand through the Anenous Pass down to Port Nolloth on the Atlantic Ocean. This road traversed empty country in the northwest Cape Province of South Africa: traditional Nama (Khoi) territory. I suddenly caught sight of a huddle of hemispherical huts nestling against a hillside.

The huts were exactly as Mr. Maxwell described them 270 years before in the letter to his eminent friend, but the young women with whom I went to talk were not dressed as they might have been then. They were fresh-faced and jolly, their delicate apricot-skinned, high cheek-boned faces smiled at me and they were dressed in blue jeans and shirts with bright floral prints or plain-coloured in cheerful pinks and yellows. On their heads they had the multi-coloured hand-knitted woollen caps seen all over southern Africa which look like tea-cosies.

I asked them what they were doing there and why they had built traditional 'Hottentot' huts of skins and cloth lashed over a framework of withies. They said that their menfolk were working as labourers on a survey for a big construction firm, Savage and Lovemore, who were going to re-build the road. As for the huts, that is how their old people always lived and when they were travelling it was still so much easier to use the old ways.

The use of such hemispherical huts by nomadic herders was universal in sub-Sahara Africa. I have seen almost identical constructions from the lands of Nilotic Samburu in northern Kenya to Bantu Himba in the semi-desert of northern Namibia to those Khoi Nama people in the Cape Province.

When travelling in the Karoo, that vast semi-desert that covers much of the Cape provinces of South Africa, one meets descendants of the ancient Khoi. They are distinctive and attractive people with their apricot-coloured faces, 'peppercorn' hair and the bright colours the women wear. It is still most exceptional to see women without a head covering.

In 1989 I was able to observe a pastoral herding group on the banks of the Kunene River in the Kaokoveld wilderness on the remote border between Namibia and Angola. A nomad family set up camp on the Angola side of the river opposite. There were two men, a woman with a baby and a boy in their party, with a dog and two donkeys to carry their possessions. Two hemispherical huts were soon set up. They had about 120 goats. It was a truly ancient scene which I watched until the sun set. They built a rough kraal and herded the kids within it while allowing the adult goats to roam about feeding on the acacia trees and riverine scrub. Their fire made aromatic smoke which drifted up to me. I watched the woman preparing food in a pot and the boy tethering the donkeys with shouts and braying, and laughter from the men. Their dog barked as the sun set and often during the night. There was a bright moon in the crystal desert air. Baboons also barked along the river cliffs, disturbed by the nomads' dog.

I was told about the nomads of the Kunene. Their herds browsed the riverine scrub for a few days in each place. During the day, the adult goats were moved up and down, while the kids were kept safely in camp. A family of half a dozen needed an average of 80 goats to survive. They met their nearest neighbours every month or two and visited the nearest trading store maybe twice a year to exchange their surplus stock for cereals, salt, sugar, clothes and tools. I watched a satisfying lifestyle there almost unchanged by modern civilisation and undoubtedly similar to a family group living in similar harsh environment at any time in the last 2,000 years in southern Africa.

The nomad family's herd of goats at the Kunene in 1989.

A Khoi father and daughter near Griquastad in 1972.

Author's photos

Perhaps of all the reasons why it is such a great a pity and a serious disaster that the Khoisan, both Khoi and San, lost their culture in the 20th century is that they had a particular mastery of the subconscious mind.

Their mastery was simple and profound. Their haunting music and trance-dances with stress-removal, religious and creative objectives are obvious. There was also a connection between their psyches and those of other large mammals with brains that work not so very differently to ours. If we could concentrate thought within the right side of our brains and eliminate language, which I presume is what happens in a trance, then we would be using our minds like gorillas or elephants.

Remove the complexities of language codes and perhaps people communicate easily with each other and, with lesser clarity, with most other higher mammals. No doubt this activity is inexplicable simply because it happens without language. What happens in meditation? Is that not moving into language-free thought: clearer and simpler, and precisely inexplicable by definition? I believe that San-Bushmen communicated psychically with other animals with ease, in a limited way. I believe that I have personally experienced it and that expert hunters, shepherds, cattlemen, animal handlers and trainers do it continually without conscious effort. Anybody who has kept pets knows something of this.

I have long enjoyed and respected the late Dr Lyall Watson's books on the edges of science, because he was not only a fellow South African of my generation with similar background, but was an academically trained zoologist with wide experience of nature in various parts of the world. In one of his later books, he wrote:

> I have been lucky. I have had the great fortune to see most of the world's wild pigs on their home ground, and have had a close personal relationship with three pigs on three different continents. ...
>
> For what it is worth, I believe that pigs have a message for us that is there for the asking, because we share so much already. We`are both products of an omnivorous upbringing, curious, dextrous and willing to explore new things. And, as a direct result of such open-minded, open-mouthed enthusiasm, we are what we have eaten. We are consequences of parallel adaptation,

genetically modified by long association with a wide range of plant chemistries that have shaped our bodies and minds.

Perhaps even more importantly, pigs and humans find common cause in the fact that we are both the recently domesticated result of a long tradition of gregarious, playful, tuneful, caring, resourceful and generally reasonable beings. We are species that can and should get along. Better perhaps than we can, or probably ever will, with any of the great apes who present us with all the problems that beset close relatives everywhere. Pigs and people are just different enough to keep each other guessing, and that looks to me like a very sound basis for any long-term relationship.

The greatest problem at the moment is a communication barrier that exists only because we persist with the stubborn notion that animals lack consciousness. The established position is so entrenched that research that even tries to prove or disprove non-human awareness, is discouraged before it can begin. This is astonishingly short-sighted when there is already sufficient evidence out there to make a very good case for animal cognition. All we lack is the courage to allow that some animals really do understand each other instead of just responding in some innate, instinctive fashion.

This acceptance is long overdue where primates are concerned. What I have tried to do here is to suggest that there are good reasons to extend that same presumption to pigs, at least as a working hypothesis. My contention is that present knowledge already shows that pigs can and do distinguish between self and non-self, and that they are able to comprehend quite complex circumstances, and to respond to them in meaningful, perhaps even conceptual, ways.

Pigs process thoughts. They understand 'if, then' situations, they apply previous experience to novel circumstances, and they interact with their environments, and with each other, as though they are conscious of consequences.

What more do you want? Pigs with wings?

Lyall Watson : *The Whole Hog*, (2004).

RESURRECTION, OR THE AFTERLIFE:

The Moon, they say, called Mantis,
sent him with life to people saying:
Go to men and tell them this -
As I die and dying live,
you too shall die and dying live.

A Khoisan poem translated by W.H.I. Bleek (1864).

Because of the reverence which Khoisan people had for the praying mantis, the messenger who brought the news of an afterlife to mankind, in South African literature and common speech it was often named, 'Hottentot god'. I have always been fascinated that the Khoisan people at the furthest end of southern Africa, far removed from the centres of modern religion and 'Civilisation', had an entrenched and central belief in an afterlife.

FOUR : *CATTLE POINT THE WAY*
Tracing the introduction of modern cattle to Africa.

The difficult task of attempting a historiography of the Khoi and the evolution of agriculture in central-southern Africa involves the prehistory of cattle. Their story, in many ways, is as complex as that of their human masters and a wry, whimsical thought has come to me at times. What would a history of mankind be like if written by a bright and intelligent, wise and thoughtful old bull?

The evolutionary track of domestic livestock is fascinating and a vital strand in a search for an understanding of Africa. The arrival of sheep in southern Africa 2,000 years ago was described in the previous chapter and the early interaction between people with herding culture and the indigenous Khoisan of eastern and southern Africa examined. The more powerful and sustained colonisation of that vast realm of the ancient roots of mankind was undertaken by Iron Age people whose agriculture was primarily motivated by cattle-keeping. Cattle were not the principal source of daily food, although the milk and derived products from their herds gave them regular protein in their diets, but cattle were at the heart of their social structures, religion and their group psyches.

From 10,000 years ago sheep, goats and cattle played an increasingly powerful role in all African societies outside of the tropical forests. The plains antelope dominated hunter-gathering society on the savannahs and cattle increasingly played a similar role, becoming more important because they were integrated into human culture in the role of property in ways that wild game never could with hunter-gatherers. Hunter-gatherers knew that antelopes and all wild animals were part of the Universe, to be studied, followed across seasonal grasslands, hunted periodically and feasted on, to be part of their trance-dances and the subject of their songs and paintings. Some animals were icons and totems to be revered. But none of them was property, they were as free as the people themselves and the plants of the soil.

Buffaloes (*Synceros caffer*) were hunted in the Sahara when it was savannah, but the African Cape buffalo has never been domesticated. There were indigenous cattle in North Africa too. Rock-art depicts the *bubolus antiqus* which archaeology tells us became extinct by about 7,000 BP, presumably hunted out of existence in a changing environment.

There is archaeological evidence from rock-art and settlement remains that sheep (ancestors of the mouflon, *Ovis orientalis*) were probably managed and selectively culled in Sahara fringes, but if they were ever bred and herded they were superseded by types more successfully domesticated previously in the Middle East and Persia. Horses have always been hunted and they were used to pull chariots in the Sahara 3,000 years ago, but they were not zebras; they were breeds introduced from the Middle-East.

The wild cattle of Europe, the aurochs (*bos taurus primigenius*) were painted gloriously by the Cro-Magnons in the Dordogne and survived the Late Stone Age hunters in northern Europe but the world's domestic cattle are well documented.

The evolution and descent of domestic cattle has been investigated with genetic and biochemical techniques which have adjusted and confirmed historical assumptions. Domestic cattle came originally from India and after selective breeding were passed on by herders from the Middle East to Egypt and onwards into Europe. These domestic cattle interbred with the aurochs.

Africans probably herded cattle in the north-east before they made their appearance around the European Mediterranean and long before the bulls were leapt over in Crete. The ability of Asian and European cattle to interbreed, even with the yak, makes their ancestry difficult to decipher, but it is interesting that they were unable to hybridise with the Asian water-buffalo and the African Cape buffalo. Presumably this was also true of the extinct Eurasian bison.

Clyde Manwell and C.M. Ann Baker in a University of Adelaide paper, *Chemical Classification of Cattle* (1980), describe the evolutionary trail and its investigation.

> The phylogenetic tree for the ten major cattle breed groups can be superimposed on a map of Europe and western Asia, the root of the tree being close to the 'fertile crescent' in Asia Minor, believed to be a primary source of bovine domestication.
>
> For some but not all protein variants there is a cline of gene frequencies as one proceeds from the British Isles and northwest Europe towards southeast Europe and Asia Minor, with the most extreme gene frequencies in the Zebu breeds of India.

Manwell and Baker's map shows that the origin of domestic cattle lies in northern India and the genetic trail moves to southern Arabia, along the watered escarpments on the edge of the Indian Ocean and up to the Middle East where a separation occurred. Two major sub-species emerged, presumably during selective breeding. Maybe this was enhanced by mutations caused by cosmic radiation at the end of the last ice-age.

Bos taurus are not humped and have a less robust dewlap. They spread into northern Africa across the Sahara and north and west through Turkey into southern Europe and eventually to the northwest limits. Over several millennia of artificial selective breeding, genetic drift and environmental pressures, *Bos taurus* evolved into the distinct races and hybrid mixes that can be found today and which have been introduced around the world during the last five hundred years of European colonial activity. In Africa the *bos taurus* types are usually known inaccurately as *zebu* and are found in the Sahel of western Africa, where they are noted for the spectacular broad stretch of their horns.

Bos indicus are humped with a heavy dewlap and do not have such long horns. They are the native cattle of India where they are confusingly also known as *zebu* because that is the origin of all the races. In Africa, *Bos indicus* are found in their purest strains in the Horn of Africa and down the eastern coast. Down the centre of the continent there are hybrid races, which are often known as *sanga*.

I have long been fascinated by the present location of these races of cattle in Africa. Domestic cattle moved with their owners and so tracing them not only shows where the original introductions took place, but indicates with some sureness how their owners migrated or where the trails of trading, clientship and symbiosis led.

After domestication and selective breeding in the Middle-East, *Bos taurus* types spread across the Sahara and, after its desertification, became confined in a West African pocket; as were their owners at the time, the Negro Congo-Nilotic core-people. It has been suggested that some of these Saharan *Bos taurus* were taken across the Straits of Gibraltar to Spain and thus introduced their particular race separately to the general spread from eastern Europe. The modern fighting bulls of Spain may be descended from this particular African sub-species with its ancient pedigree. P.H. Starkey wrote in *World Animal Review, v 50* (1984):

> It is thought that Hamitic long horn cattle entered Egypt around 7,000 years ago, and their owners migrated across North Africa about 5,000 years ago. From north-west Africa, some of these cattle moved northward into Europe, and some moved southward into west Africa.

It is generally accepted now that domestic cattle of the long-horned type, *Bos taurus*, were introduced to the Sahara region at about 7,200 years ago, at the beginning of the Early Pastoral cultural period.

Bos indicus, the closest domestic race to the original Indian progenitor, were introduced from the Yemen, in southwest Arabia, to Ethiopia by sea, probably 4-3,000 years ago. From there they spread southwards with Cushitic nomads and also became mixed with *Bos taurus*, which had already spread across the Sahara to the southern Sudan and the savannah parts of today's Central African Republic and Uganda, taken there by Nilotic herders.

These *sanga* hybrids may have been introduced to the aboriginal Khoisan people who became the Khoi herders of southern Africa. Certainly they went south down the central savannah trail in the first millennium A.D. and have particular relevance when tracing Iron Age migrations into southern Africa.

Sanga cattle were encountered by Portuguese ocean explorers in 1488 and 1497 on the South African coast. They thought that the Khoi herders, still economically in the Late Stone Age, were primitive, but they admired their cattle which no European had seen before. The *Diário da Viagem* of Vasco da Gama's voyage says:

> ....and there [Mossel Bay] we traded a black ox for three bracelets. We dined off this on Sunday, and it was very fat; and the flesh of it was as savoury as that of Portugal.
>
> ..... The oxen of this land are very large, like those of the Alentejo, and very marvellously fat, and very tame; they are castrated, and some of them do not have horns. And the negroes fit the fattest of them with pack-saddles made of reeds, like those of Castile, and on top of the pack-saddles some sticks to serve as litters, and they ride on top of them. They thrust a cistus stick through the nostrils of those cattle they wish to barter and lead them by that.

As an aside, it is interesting that the early herding culture of the Sahara also developed a tradition of using oxen as pack animals. Using pack-oxen and the riding of them, even in warfare, was common practice from southern Europe to the end of Africa.

After the Dutch colonised the Cape, the *sanga* cattle of the Khoi were developed by breeding into the pedigree Afrikander with its rich red colour, small horns, large hump and swinging dewlap.

Late Iron Age cattle in KwaZulu-Natal have been identified as *sanga*. In that part of Africa, the Zulus used particular breeding technique in modern times and when Europeans first settled there, they found that King Shaka had a military organisation in which different regiments were responsible for herds with distinctive colouring and hide patterns. Their six feet tall battle-shields were made from these 'regimental' herds so that each regiment in formation presented an awe-inspiring, uniform appearance. This refinement of selective breeding was perfected by King Cetshwayo whose personal herd was pure white. A remnant of this royal herd was husbanded by the KwaZulu Cultural Museum at Ulundi in the 1980s.

The Food and Agricultural Organisation of the UN published a paper by E.P.Cunningham and O.Syrstad in 1987, *Crossbreeding bos taurus and bos indicus for milk production in the tropics* advocating procedures which 'primitive' native Africans had already discovered through trial and error probably at least 3,500 years ago! The Preface states:

> In recent decades there have been many attempts to improve the productivity of indigenous dairy cattle in

developing countries by crossing them with temperate breeds. Results have been variable with well-known examples of both success and failure. During this process much has been learned about the percentage of temperate breed genes which can be successfully introduced and also about lost adaptation of the crossbreds if the percentage of indigenous breed genes is too low. It is now clear that under most lowland tropical conditions the most productive dairy animal is a crossbred between Bos indicus and Bos taurus.

Camels replaced cattle, sheep and goats throughout the Sahara and in the more arid parts of the Sahel as desertification progressed. Sheep and goats were able to resist some tropical diseases which afflicted cattle and often preceded them southwards, or could be kept in small numbers by mixed farmers where cattle could not survive. The fat-tailed sheep of Persia arrived in southern Africa ahead of cattle, about 2,000 years ago.

I consider that the dominating concept of property is central to civilisation, which might be considered to be a universally instinctive trait of mankind today, and it influenced the early pastoral herders. Instead of managing and breeding their animals for meat, herders developed an economy of feeding off their milk and processed dairy products. Some added blood to their diets by carefully calculated regular bleeding. The modern Masai of East Africa are the tribal people most quoted in this respect and numerous books and TV documentaries feature the charismatic Masai. Bleeding was not only an African practice and Mongol nomads of the central Asiatic plains got their protein nutrition by bleeding their horses.

Surplus domestic cattle and small stock were killed and eaten, especially goats and sheep, but this was usually practised for a ritual, a celebration, a particular feast or entertainment. Livestock became precious property, used for essential barter trade and prestigious presents, the currency of bride-purchase and the tribute of client clans and tribes. They were not mere walking larders to be casually slaughtered for daily food.

Those later Iron Age people who developed mixed agriculture in the tsetse-free healthy zones of eastern and southern Africa revered their cattle as much as the nomadic herders did and, despite their bulk nutrition coming from cereals and vegetables, many kept to a cattle-cult. Semi-sedentary people, living almost entirely from gathering, cultivation and fishing kept a few domestic cattle or goats where they could survive disease for prestige, rituals and their milk. Livestock became a powerful common currency of Africa: the link between herders, mixed agriculturalists and farmers. Increasingly, it became the trading link between them and the miners and metallurgists as iron became dominant in society.

When population pressures increased in optimum mixed agriculture geographical zones during the last few centuries, cattle-raiding leading to prolonged inter-tribal feuding and warfare became endemic. Conflict between the Kamba and Masai in East Africa and the Zulu and neighbouring Nguni and Sotho clans in South Africa are typical examples. King Shaka of the Zulus would reward, train or entertain his army by sending them off on cattle raids. The Nguni offshoots, the Ngoni, in East Africa were notorious for their cattle-raiding which disrupted previously peaceful societies. The first serious conflicts between European settlers and native Africans in South Africa in the 18th and 19th centuries was always over cattle or their foraging ranges.

Modern African herders captured sentimental imaginations with their 'nobility', cattle-cults and heroic warrior traditions, but the produce of the soil was vital to life and much ritual was also applied to cultivation. Rainmaking and the celebration of harvests was a central role of chiefs, or priests when there was a separation of religious and executive functions. Precious animals were sacrificed to promote the rains in the fickle climate of the eastern and southern African savannahs, because without rain farmers starved and so did domestic livestock.

The Congo rainforest waxed and waned in wealth and distribution as climate cycles demanded. It has been an impenetrable island in central Africa for millennia around which people who were not adapted to subsistence within it have had to migrate. Cattle could not be moved through it and extensive cultivation could not take place without clearing the great trees. The considerable effort was seldom worth it until iron tools were readily available, and difficult and tedious thereafter. Africa was big enough for the earliest cultivators and herders to seek other pastures rather than to force their way through the rainforests. The explosion of population along the southern fringes of the Sahara in the first millennium BC, when agriculture with iron tools and advanced herding technique became generally practised, squeezed people around the Congo forests and southwards along the ancient route of the Great Rift Valley.

No matter that the Sahara acted as a filter, delaying change by hundreds of years, in West Africa the influence of Nilotic civilisations was powerful. Organised societies, tribal structures and the evolution of many dialects and languages proliferated. Territories became increasingly demarcated and warfare occurred as populations grew and the Sahara desert expanded, creating compression and structural changes in society.

With the resources of an Egyptian heritage onto which to build new ideas and adjustments, the people of West Africa evolved complex and highly organised national feudal states and empires. Mungo Park described their remnants and their cultures, still complex, during his travels in the early 1800s. These kingdoms and empires drew their military strength from the traditions of nomadic people who kept to a cattle-cult and revered domestic animals. Their armies included cavalry mounted on horses introduced from Arabia as well as camel corps. Today, titular chiefs and Emirs in Northern Nigeria continue to hold ceremonial 'durbars' featuring processions of cavalry and wild charges of men on horseback.

Walled towns were built for defence in territorial war and to combat Arabic slave-raiding, with warrens of square houses within. Ruling these towns and the agricultural fields and range-lands about them, a balanced system and tradition of dynastic inheritance was generally established which had possible Egyptian origins. There was a ruling king at the head who symbolised the nation and its ancestors, but descent proceeded down the female line. It was a matrilineal system and this was applied to property in the general populace. The king's senior or most influential wife at his death chose a suitable daughter to be queen and her choice of spouse became the next king, and so on. It was a workable system of checks and balances which lasted for millennia in an introverted semi-urban society within harsh lands. Matrilineal ways of determining inheritance have persisted not only in West Africa, but amongst those people who migrated from there most directly into southern Africa in the first millennium after Christ.

The earliest appearance of agriculture southwards of the swamps and floodplains at the head of the White Nile, particularly around Lake Victoria and along the two branches of the Great Rift

Valley, seems to have been about 5,000 years ago, before metals were available. The resident hunter-gatherers and fishermen became clients and were absorbed in time. Coincidentally, further east, Afro-Asiatic people, presumably Cushitic, were moving south out of Somalia, southern Ethiopia and the eastern Sudan.

*   *

A picture can be drawn with the crudest of strokes at about 4,000 years ago. A concentration of mixed farmers and herders with *Bos taurus* cattle and horses grew in the western Sahel as population and climate pressures dictated. Those that entered the fly-infested forest margins lost their cattle and mixed with the native forest-dwellers. In the centre, farmers who learned to cultivate millets and sorghum, settled near lakes and rivers where they could fish for protein. Until the climate changed sufficiently and enough bush was cleared by these farmers, exotic livestock from the shrinking Sahara grasslands was endangered. From the north of the Rift Valleys and the Kenya Highlands, Neolithic herders moved south and Cushites came from Ethiopia, keeping to healthy ranges where there was either perennial water, or water could be found by moving back and forth with the seasons.

The north-eastern cattle-people had *Bos indicus* strains which were transported across the narrow strait of Bab el Mandeb from the wealthy and prosperous kingdoms which developed on the escarpments and highlands of southern Arabia. It was those kingdoms which became great traders, supplying frankincense to the Middle-East and providing entrepôts for the growing coastal trade between India and Egypt. Ethiopia and the Yemen became linked by trade and the exchange of people, animals and goods. The geography was similar and the Horn of Africa had access to the highly developed nations of the upper Nile, the great cities of Meroe and Axum, and onwards.

Nubian kings ruled Egypt itself briefly during the first millennium BC and there was trade connecting the Mediterranean and Ethiopia up the Nile as well as along the Red Sea. Ethiopia and the southern Sudan may have been overshadowed by the marvels of Lower Egypt, but the nations and cities that grew there had direct impact on eastern and southern Africa, the core-lands of all mankind, through diffusion of the cattle-cult.

Traditional native Nguni cattle with humps and dewlaps preserved for posterity in South Africa. Photo from the Internet

Local Afrikander breed commonly used as draft animals into the 1960s.

Photo by the author in 1949

FIVE : *INDIAN OCEAN SEATRADERS*
The great Indian Ocean trading system.

> The story of what the ancients accomplished on the sea has never been put between the covers of a book. A few episodes have been dealt with so often, in handbooks and histories, that they are as familiar as Caesar's assassination. But once off these well-trodden paths, the searcher for information is forced to make his way through a miscellany of scholarly publications, more often than not articles in obscure journals, in a variety of languages, and he will find that some topics have never been treated at all.
>
> Lionel Casson in *The Ancient Mariners* (1959)

I wrote in my book, *Two Shores of the Ocean* (1992), about my frustrations when attempting a study of Indian Ocean history and pre-history in 1970:

> I learned that the Indian Ocean was routinely navigated for about 2,000 years before Columbus crossed the Atlantic, but the scale of this human activity eluded me. I wished to read about it but found nothing to read.

Thankfully, there have been improvements in the last forty years, but many old myths and misconceptions are deep-rooted.
Stephen Oppenheimer in his book, *Out of Eden, The Peopling of the World* (2003), devotes much space to the thesis that one of the

two main routes for the genetically-proven migration of Middle Stone Age modern Africans to Arabia and thence along ocean shores into Asia was across the Strait of Bab-el-Mandeb. This strait is at the Horn of Africa where the Red Sea narrows before opening into the Indian Ocean.

Today it is a mere thirty kilometres wide and has the island of Perim within it to reduce the actual sea distance. Geological evidence shows that this part of the Red Sea was unstable during the last two million years and more, and coincidentally succeeding ice-ages and warm periods caused the ocean levels to rise and fall. At about 80,000 years ago, when there was a major migration from the Horn to Arabia, the strait was open and the Red Sea was flooded. Nevertheless it was shallow and there were islands and reefs which are not visible today, so people at that time must have used rafts and canoes to cross over. If this was their method of migrating, they must have had prior knowledge of some kind of seafaring however primitive.

The modern people who migrated out of Africa traversed the shorelines of Asia, moving up river valleys when they found them, and reached Indonesia probably before 75,000 years ago. After the Toba volcanic winter the oceans were lower, but people nevertheless crossed the channel from Timor to Australia where the sea was too deep to dry out.

There has been speculation for some years that the first migrants from Asia to North America moved by sea, island hopping, long before the mass movements of 'Clovis People' across the dried-out Bering Strait 18,000 - 12,000 years ago. This was based on finds of stone implements found in various places from Canada to Chile, but their provenance was always subject to dispute. However, this speculation was supported in 2005 by the announcement that human footprints had been discovered in volcanic ash dated to 40,000 years ago near Puebla in Mexico. Dr Silvia Gonzalez of John Moores University in Liverpool and her team found the footprints in volcanic ash in 2003, but had to wait until July 2005 to announce her results when the date had been confirmed by techniques at Oxford University. The clear inference is that between eighty and forty thousand years ago, people had migrated from Siberia to Mexico and this had certainly occurred along the Pacific seacoast and possibly by island-hopping from place to place and from time to time.

Another persistent legend is that European Neolithic people, known as the Salutrians, migrated from Scandinavia or Britain around the northern rim of the Atlantic to North America and eventually settled in the Great Lakes region. Stone tools attributed to them more resemble European Neolithic than those of the migrants from Siberia. There is evidence that some native North Americans have 'European' genes, although whether these genes were acquired thousands or hundreds of years ago is not clear. What is of interest is that research into this speculation shows that it was perfectly possible for Neolithic people to have migrated along the edge of frozen seas using simple canoes constructed from animal skin stretched on wooden frames as built by modern Inuit-Eskimos.

Modern Eskimos went on long hunting journeys and easily survived from the fruits of their hunting and fishing and, whenever bad weather menaced them, they landed on ice floes and sheltered under their canoes. The technology used by these modern hunter-gatherers was certainly available to Neolithic peoples all around the northern Pacific and Atlantic rims during ice-ages when migrations occurred. The ability of people to have migrated from both Siberia and western Europe to North America at any time during ice-age years cannot be in doubt. Since modern people did not reach the Atlantic shores of Europe until about 40-50,000 years ago, a migration of Salutrians to north America from Europe must have occurred after then.

There has also been persistent scholarly speculation that West Africans crossed the Atlantic to found an Olmec colony in Central America variously between 10,000 and 2,500 years ago. I have no doubt that Africans could have navigated the Atlantic using the west-flowing Equatorial Current, but I wonder if these migrations will ever be satisfactorily proven. My correspondent, Harry Bourne, has devoted much scholarly effort into researching ancient navigation in the Atlantic and his material is available on the Internet.

I believe there can be no doubt at all that there were regular island-hopping seafarers and fishermen in boats off the East African coast long before 40,000 years ago. There is nothing about a large sea-going dugout canoe which is still being used today all around the rim of the Indian Ocean, on the West African coast and in the Pacific, that would defeat the ingenuity of Stone Age people many thousands of years ago. Today they are still hand-made with steel axes and adzes and stone tools would have taken longer. But Late Stone Age people certainly cut down trees and if they could do that, they could fashion a canoe identical to any that are presently in use. One commodity that pre-civilisation people always had in quantity and which they never bothered to measure with any anxiety was time.

It is surely certain that Early Stone Age people, were paddling logs or primitive canoes and rafts a half-million and more years ago? We have to remember that a *Homo erectus* community had crossed the open sea to Flores about 800,000 BP as evidenced by their stone tools. *Homo erectus* crossed the Strait of Gibraltar to Spain before 700,000 years ago. Polynesians were observed swimming and fishing without boats, using simple personal floats, far out to sea in the 1920s. Probably *Australopithecus* hominids, like modern children in a swimming pool, held onto floating objects and kicked themselves along when crossing rough or dangerously long stretches. All people with access have played in the water since childhood and harvested it for food, so floats must have been used as early as stone hand-axes, carry-bags and fish-spears. This is at the core of the Seashore Hypothesis.

Thor Heyerdahl, famous for the Kon-Tiki voyage early in his career, promulgated his belief in the spread of Late Stone Age culture across the oceans. Heyerdahl attempted to prove particular theses in which he believed by carrying out spectacular voyages in careful reproductions of ancient sailing craft. Heyerdahl's much publicised first voyage across the Pacific in *Kon-Tiki* from Peru and the simplistic thesis he promoted then was overtaken by other evidence that Polynesia was colonised by people from southeast Asia and the substance of his beliefs and his reputation suffered. Attempting to argue his corner against accumulating, irrefutable contrary evidence increasingly damaged his reputation. However, I cannot doubt that some ancient South American people sailed the Pacific in balsa rafts like the *Kon-Tiki*.

Heyerdahl carried out a lifetime of work on ocean-voyaging by Late Stone Age peoples right around the equatorial belt as well as the maritime excursions of the grand civilisations. He sailed a reed vessel from the Persian Gulf to the Horn of Africa. He and a professional team got permission to investigate Late Stone Age and first millennium AD seafarers, with intercontinental megalithic cultural links, in the Maldive Islands of the Indian Ocean long before those enchanting islands were opened to tourists. I was particularly charmed by one fact they established; that the Maldives were a source of cowrie shells traded all around the western Indian Ocean and which have had a value far exceeding any intrinsic worth; a form of currency over vast areas of Asia and Africa until the 20th century. His more mundane yet important researches have had far less attention than 'Kon-Tiki' and his Pacific obsession.

* *

Lionel Casson in *The Ancient Mariners* (1959) wrote:

> "Bringing of forty ships filled with cedar logs." So wrote an ancient scribe in listing the accomplishments of Pharaoh Snefru, ruler of Egypt about 2650 B.C. This handful of words brings one across the threshold into the period of history proper.

A characteristic of three centres where earliest civilisation emerged, Egypt and Mesopotamia and northwest India, is their dependence on great rivers. Not only did the Nile, Tigris, Euphrates and Indus provide ideal environments in which to perfect agriculture and build complex urban societies, but they provided marvellous highways. Boats and ships could ply the rivers with ease and carry quantities of goods with little effort. Sailing ships were invented and in widespread use by Egyptians on the Nile before 5,000 years ago, proven by pictures on pottery. A University of Pennsylvania expedition led by David O'Connor in September 1991 discovered a fleet of sea-going vessels ritually buried in the desert near the temple complex of Abydos, some eight miles from the present course of the Nile. These ships were buried about 5,000 years ago. An ancient ship, also dated to about 5,000 years ago is preserved at Giza.

Density of urban populations within those early civilisations reached levels comparable to modern industrial nations through the richness of their agriculture, metallurgy and the centuries-long stability and order of the state. With masses of people available for public works every year after the harvest was gathered and before the next sowing season came around, the palaces, temples and monuments which were created exceeded in wealth and sheer size anything that later European empires would attempt. All along the northern sub-tropical zone where these favourable conditions of seasonal river flooding or heavy rainfall occurred, there were similar explosions of people and material wealth. Separated by millennia or oceans, the same pattern emerged in Central America, Egypt, Mesopotamia, India and southeast Asia. But Egypt and Mesopotamia were first.

Metals provided these first great civilisations with another spur. Copper tools enabled precise carpentry which enabled efficient brick making, house construction and shipbuilding. When alloying was discovered and bronze was invented, precise stone masonry became practical, megalithic buildings were erected and weapons of war greatly improved. And when the mysteries of extracting iron from their ores and the smelting of steels were mastered, then the material advance of mankind in the ambience of those civilisations was limited only by mankind's ingenuity in developing machines to speed work.

The Egyptians progressed to the limits of the development of machinery by 5,000 years ago which prevailed, more-or-less, until the Renaissance and succeeding Industrial-age in Europe. Greeks, Minoans, Chinese, Indians, Persians or Romans improved this or that technique, refined knowledge and experience in one direction or another, but the foundations were laid, the bases were all there. The next jump forward could only happen with a greater knowledge of physics and chemistry and the world had to wait for the likes of Galileo, Descartes, Isaac Newton and Michael Faraday. Even so, modern machines and the harnessing of electricity and complex chemistry has maybe altered only the *speed* of our lives. Has it improved the daily quality? Can anybody be sure that a fisherman, peasant farmer or petty trader in Egypt, Iraq or Pakistan in 2,000 AD enjoys life more than his equivalent in 2,000 BC? I believe that his quality of life has declined.

The Egyptians had easy sources of copper, zinc and tin in the Eastern Desert between the Nile and the Red Sea. There was also some gold in those hills. On the Nile, nearest the mines, a cluster of temples, palaces and cities grew over the centuries at Abydos, Dandara, Karnak, Luxor, Naqada, Thebes and Idfu. Today, the same hilly country along the Red Sea is still a source of some copper, gold, zinc, tin, chrome, manganese and asbestos. The riches of Egypt were indeed almost magical.

Not only was the land of Egypt particularly blessed, but it had easy access to its equally blessed neighbours in Mesopotamia. Innovations, new products and ideas were introduced from one to the other and traded as profitably as goods. Improved grain and domestic animal species were introduced to Egypt from the Middle-East and these fuelled Egypt's agriculture and population explosion. The science of iron-smelting came originally from Anatolia, brought by the Hittites, but Egypt had much to offer; particularly the products of its efficient farms, linen industry, metal mines, exceptional scholarship, culture and the enormous stability of its civilisation at a geographical cross-roads.

Intellectual ideas as well as agricultural and iron technology were exchanged between Egypt and the Middle-East and Mesopotamia to be exploited by the industry of the Nile peoples. So too did ocean voyagers. As Roland Oliver wrote (1991):

> The Wadi Hammamat led not only to the mines, but also to the Red Sea, whence came long-robed, bearded strangers, well-depicted in contemporary paintings and carvings, who were very different in appearance from the linen-kilted, clean-shaven Egyptians, and who used ships with tall, backward-curving prows, quite unlike those which plied on the Nile and Mediterranean. The strangers were certainly from Mesopotamia, probably from the seaport of Susa on the Persian Gulf.

The caravan routes connecting the three great centres of wealth and power, Egypt, Mesopotamia and India, were long, hard and dangerous. There were deserts and mountains between them, inhabited by wild tribes with brigandish morals. The Indian Ocean became the best highway, especially for the transport of valuable cargoes which benefited most from speedy transport. The longer a merchant's wealth is tied up in transit, the slower his cash turnover and the less his profit over time. If you can buy and sell a parcel of goods in a year rather than three or four, you make three or four times as much profit in a particular period and your rate of reinvestment and expansion compounds that much faster. Sumerians, Babylonians, Assyrians, Egyptians, Phoenicians, Jews, Persians and Indians understood this simple rule of commerce and growth.

The Suez land-bridge led straight to the rich Jordan valley, and via Syria to Anatolia and the upper reaches of the Tigris and Euphrates. These were practical routes for armies and caravans and much used by both, back and forth. Arabian overland trails were used, of course, typically to the Yemen for aromatic resins, especially when the camel came into general use, but Egypt had a special position in relation to seaborne trade. Egyptian ships could sail the Mediterranean from ports in the Nile delta. From the Red Sea there was access to the Indian Ocean from ports close to those mines in the Eastern Desert and the centres of power at Thebes and Luxor. Hormos, near today's Quseir, and Berenice were the Red Sea ports. The Nile itself was the highway into the interior of Africa and the wealth of Nubia. Egyptians became great traders.

After establishing efficient agriculture and the technology for constructing cities and monuments, the path to magnificence was trade. It is the only way to exploit your own surplus goods and to obtain peacefully the surplus goods of other societies. Trade is a 'driving force' of civilisations and the universal natural laws apply. Considerable effort was devoted to the development of shipping.

During the second millennium BC, between 4-5,000 years ago, Sumeria, Egypt and the Indus Valley civilisations thrived and were interconnected by trade. Owing to a lack of forests of large trees, reed boats were developed in the Persian Gulf for fishing and coastal trade and improved to sail further afield. Wood was always one of the mainstays of this regional trade, and remains so today. The Indus Valley Civilisation, with capitals at Harappa in the Punjab and Mohenjo Daro in Sind, built fine port cities and fleets of wooden ships. No doubt the Indus River encouraged the construction of barges and canoes which led naturally to sea-going vessels as that civilisation expanded in trading activity and technology. At Lothal, in the Gulf of Cambay, a dock with gates and sluices was built so that ships could lie afloat alongside the loading quays at all states of the tide. Egyptian ships sailed the Levantine coast in advance of the great Phoenician seafarers who followed. The southern entrance to the Red Sea achieved importance and the foundations of the trading kingdoms of the Yemen in southwestern Arabia were laid. At the entrance to the Persian Gulf, trading cities grew in what was to be called Oman by later Arab inhabitants.

Camels were domesticated before 3,000 years ago and the long lines of their caravans began moving across the lands of Arabia and North Africa where the deserts were spreading as climate changed. Camel caravans carried bulky cargoes: salt, minerals, timber, cloth and grain and they could equal the speed of ships on direct overland routes where there were no geographical or human barriers. The frankincense trade between southern Arabia and the Middle-East was carried by camels. But ships continued to have a principal part to play. Ships crossed from southern Arabia to the Horn of Africa which shared in the frankincense boom. They carried valuables which the great nations increasingly desired from each other and could sail directly across the oceans between India, the Persian Gulf and the Red Sea; routes which were not easy for caravans because of mountains, hard country and wild men. Ships carried refined gold, silver, lead, tin, iron and copper, ivory, spices, fine timber, works of art and crystalline rocks and stones. Exotic plants, animals and slaves were passengers.

Iron was first exploited by the Hittites and the technology was obtained by the Philistines and carried onwards to Palestine and Egypt. From Greece and Macedonia the Iron Age surged across the Mediterranean. From Egypt it crossed the Sahara, although it is generally accepted that iron may have been smelted in the Sahel region and around the Great Lakes before it was adopted in Egypt. The Assyrians in northern Mesopotamia also obtained iron technology from their neighbours in Asia Minor and began flexing their muscles. From Mesopotamia, the technology travelled eastward through India. Copper and its alloys of bronze and brass continued to be widely used, especially for ornaments and decorative ware, but iron replaced them for tools and weapons. Another enormous technical jump was occurring.

By the beginning of the first millennium BC, two invasions from Asia Minor, the Hittites and the mysterious 'Sea People', had broken up decaying and corrupt regimes on the Levantine coast. Egyptian authority outside the Nile Valley was lost and the Israelites under Joshua conquered the rich and fertile Jordan Valley. King David expelled the Philistines from Palestine and the golden age of Israel followed.

At the head of the Gulf of Aqaba, where the Great Rift Valley is near its beginning, splitting Sinai from Arabia, an industrial city was built in King Solomon's time, in the 10th century BC. It was called Ezion-Geber, near Elat where European tourists today flock for enduring sunshine beside the sea with its coral reefs. Metallurgy flourished there within easy access of copper and iron-ore mines and archaeologists found a system of efficient furnaces in a well-organised town. Dr Nelson Glueck of the American Schools of Oriental Research was excavating and publishing the results in 1938 and 1939.

King David is credited by some with wresting the secrets of iron from the Philistines and the good fortune of abundant ores along the Rift in southern Israel enabled the growth of trade in metals, but it was the Phoenicians who provided practical know-how. Israel in Solomon's time became an important regional power. Lacking knowledge of the sea, an alliance was made with the Phoenician King Hiram of Tyre and the combined Phoenician and Israelite fleets commenced ocean trading. Hiram also sent masters of metallurgy to work the mines and smelters.

Solomon's alliance with the Queen of Sheba, controlling the entry to the Red Sea from her capital, Marib, in the Yemen had to follow. These political activities were all aimed at mutual advantages in trading within the embrace of the Indian Ocean.

From the First Book of Kings, 9 & 10, (*New International Bible*):

> King Solomon also built ships at Ezion-Geber, which is near Elath in Edom, on the shore of the Red Sea. And Hiram sent his men - sailors who knew the sea - to serve in the fleet with Solomon's men. They sailed to Ophir and brought back 420 talents [about 15 tons] of gold, which they delivered to King Solomon.
>
> When the Queen of Sheba heard about the fame of Solomon and his relation to the name of the Lord, she came to test him with hard questions. ...
>
> And she gave the king 120 talents [about 4 tons] of gold, large quantities of spices and precious stones. ...
> Hiram's ships brought gold from Ophir; and from there they brought great cargoes of *almugwood* and precious stones. ...
>
> King Solomon gave the Queen of Sheba all she desired and asked for, ... Then she left and returned with her retinue to her own country.
>
> The king had a fleet of trading ships at sea along with the ships of Hiram. Once every three years it returned carrying gold, silver and ivory, and apes and baboons.

There have been many attempts to locate Ophir and these sparse verses from the Bible have been analysed over and again. King James' Bible states that "apes and peacocks" were brought back, which would confirm India, whereas the New International Bible translates the Hebrew as "apes and baboons", which proposes Africa. The critical reference is the length of the voyages, which can only mean that these fleets were using the Indian Ocean monsoon system: allowing one year on the outward journey, a year to explore, trade and refurbish and a year to return. The destinations could then be as far as southern India or eastern Africa , or both, calling on merchants at the ports in the Red Sea inwards and outwards: requiring the friendship of the Sabaeans of the Yemen and prompting the Queen of Sheba's "hard questions" of Solomon's motives.

The Sabaeans are of great importance for the Indian Ocean sea trade. For a thousand years or so they ruled the southwestern corner of Arabia, and controlled the southern Red Sea and its outlet to the ocean. They were seafarers and advanced cultivators, building famous dams and irrigation works and the fabled agricultural terraces on mountainsides, still used today. The Sabaean dynasty was superseded by that of the Himyarites about 115 BC, but that was a political change and did not necessarily affect the activities of the merchants and sailors who continued to trade throughout the region.

According to the Bible, King Solomon of the Jews and his Phoenician sailors imported gold and ivory, scented woods and exotic animals. East Africa south of Ethiopia has no large or easy sources of precious metals, certainly not near the coast, so 'Ophir' was probably in Arabia or Ethiopia, or both. The Queen of Sheba came from her stronghold in southwest Arabia with gold for Solomon. No doubt the Sabaeans exchanged gold for the manufactured metal tools and weapons of those furnaces at Ezion-Geber. When Alexander conquered Persia enormous quantities of gold were found in the great palaces of that empire. That gold had been mined somewhere in the Middle East and Indian region.

Although of a considerably later date, there is historical evidence of the gold trade with Punt, which is likely to have been known as Ophir in previous centuries. Cosmas Indicopleustes was an Egyptian monk, possibly travelling as a trader, who recorded his voyage from Egypt to India, during which he stopped at Adulis on the Red Sea in 522 AD. G.A.Wainwright in the journal, *Man* (June 1942), published this account as translated by J.W.McCrindle (1897):

> The country known as that of Sasu is itself near the ocean, just as the ocean is near the frankincense country, in which there are many gold mines. The King of the Axômites accordingly, every other year, through the governor of Agau, sends thither special agents to bargain for the gold, and these are accompanied by many other traders – upwards, say, of five hundred – bound on the same errand as themselves. They take along with them to the mining district oxen, lumps of salt, and iron, and when they reach its neighbourhood they make a halt at a certain spot and form an encampment, which they fence round with a great hedge of thorns. Within this they live, and having slaughtered the oxen, cut them in pieces, and lay the pieces on the top of the thorns, along with the lumps of salt and the iron. Then come the natives bringing gold in nuggets like peas, called *tancharas,* and lay one or two or more of these upon what pleases them – the pieces of flesh or the salt or the iron, and then they retire to some distance off. Then the owner of the meat approaches, and if he is satisfied he takes the gold away, and upon seeing this its owner comes and takes the flesh or the salt or the iron. If, however, he is not satisfied, he leaves the gold, when the native seeing that he has not taken it, comes and either puts down more gold, or takes up what he had laid down, and goes away. Such is the mode in which business is transacted with the people of that country, because their language is different and interpreters are hardly to be found. The time they stay in that country is five days more or less, according as the natives more or less readily coming forward buy up all their wares. On the journey homeward they all agree to travel well-armed, since some of the tribes through whose country they must pass might threaten to attack them from a desire to rob them of their gold. The space of six months is taken up with this trading expedition, including both the going and the returning. In going they march very slowly, chiefly because of the cattle, but in returning they quicken their pace lest on the way they should be overtaken by winter and its rains. For the sources of the river Nile lie somewhere in these parts, and in winter, on account of the heavy rains, the numerous rivers which they generate obstruct the path of

> the traveller. The people there have their winter at the time we have our summer. It begins in the month Epiphi of the Egyptians and continues till Thôth, and during the three months the rain falls in torrents, and makes a multitude of rivers all of which flow into the Nile.

Apart from identifying a regular source of gold during this historically neglected period after the decline of the Roman Empire and before the emergence of Islam, the method of trading is precisely similar to the often-quoted way used by Carthaginians on the Atlantic west coast of Africa in the years BC. I find this remarkable and suggests that traders of the Mediterranean nations had a common base of trading know-how in use with sub-Saharan Africa for many centuries. The trading depot described in the account by Cosmas is identified by Wainwright as Fazogli, spelled Fazughli on modern maps. It was a place on the Blue Nile at about 11°16'N.34°45'E which is today on the banks of the Roseires Reservoir. Agau, whose governor the King of Axum consulted, lay on the shore of Lake Tana (11°35'N. 37°23'E.), the source of the Blue Nile.

India had ivory and monkeys, but if I were a captain facing a journey of several months I would be most reluctant to ship live monkeys needing water and the same kind of food as my crew earlier than I had to. Exotic animals and their skins were probably picked up around the Horn of Africa. Ivory was also easily available there. I think India was where they picked up jewellery, fine cloth and other quality manufactured goods using the gold of Ophir to pay for them. They used some of their Indian goods in exchange for the animals and ivory on their way up the Red Sea. And there were Egyptian and Nubian merchants from whom they could take commissions and offer trade to.

Indians are often bypassed when considering ancient seafarers and ocean traders, maybe because in European historiography, after 1500 AD, local navigators in the western Indian Ocean were principally Arabs and Islamic. The Indus Empire was so advanced in seafaring that they built fine ports for ocean-going vessels even to the extent of devising docks in which ships could lie undisturbed by the tides. Indian ships were trading with the Persian Gulf, Africa and Arabia from at least the middle of the first millennium BC and possibly earlier. Certainly there is evidence of

Indian goods reaching Egypt and East Africa at that time. Indian trading with Indonesia and further east was huge in quantity and value for many centuries. In the tradition which I have repeatedly ascribed to traders elsewhere, knowledge and culture proceeded with goods. Hinduism and Buddhism became deeply fixed in Indonesia and southeast Asia. Buddhism moved to China and through that vast country to become entrenched in Tibet on the other side of the Himalayas from where it originated; an amazing circling migration of knowledge and tradition.

Apart from the activities of Indian seafaring merchants, India was always a convenient halfway point between the two Indian Ocean monsoon systems and the nations faced by them. Indian ports of the Malabar and Coromandel Coasts served as peaceful entrepôts for traders of all nations, from as far as China, for many centuries. The industry and commerce of the Middle East and southern Asia exceeded anything that existed in Europe until the Industrial Age following the Renaissance.

Alexander extended Greek influence throughout the region and Greek skills and enterprise were added to the trading and seafaring networks. When the Romans conquered Palestine and Egypt, they inherited control of a golden domain of trading and commerce and extended it to encompass a network which stretched from Britain to China.

The Nabateans were of importance during the later first millennium BC. They were great traders and their city of Petra in Jordan proves their wealth and success based on frankincense from the Yemen and the Horn of Africa. During the Ptolemaic period Nabatean piracy was rife at the entrance to the Red Sea and hindered Egyptian and Sabaean navigation. It was the Romans who cleared the area of most of the pirate strongholds, especially that based in Aden, and made the Indian Ocean free again.

The oft-quoted motto, "There is always something new out of Africa", is more correctly: *Unde etiam vulgare Graeciae dictum 'semper aliquid novi Africam adferre'*. Which is translated: "Whence it is commonly said amongst the Greeks that 'Africa always offers something new'". This is a significant enlargement of Pliny's more common, simplified quote. He is acknowledging that the Greeks, before the Romans, had extensive knowledge of the African trade.

*   *

It has been suggested by some writers or historians that East Africa, south of the Horn, became a part of the Indian Ocean trading system almost by chance, when sailors arrived there because they were off course. This is clearly incorrect. There had been movement about that part of Africa for several millennia and knowledge, however sketchy, was known in local legend if not by contemporary overland traders and travellers. In any case, Chami's archaeology proves trading contact by sea as far south as Zanzibar early in the first millennium BC, and possibly earlier.

I believe that the ivory trade promoted contact with East Africa in the first millennium BC. Israelite and Phoenician fleets of 3,000 years ago were engaged in a three-cornered trade. They carried linen cloth from Egypt, fine timber from the Lebanon whose forests were being steadily demolished, and raw copper and metal manufactures of iron, copper, bronze and brass from the furnaces and smithies of Ezion-Geber for as long as there was sufficient local wood for smelting.

From southern Arabia they could pick up frankincense. In India there were precious metals, ivory, spices, fine cotton and silk cloth and new food plants. On the coasts of the Horn of Africa there were frankincense and other aromatic gums, ivory, exotic animals and their skins. But, further south, ivory and animal products may have been a better bargain for shrewd traders prepared to spend a few more weeks at sea. The legendary shrewdness of Jewish traders and their allies was forged in the Levant but also around the western Indian Ocean.

During the northern hemisphere summer, May-November, the Indian Ocean monsoons blow from the south and during the winter, December-April, they blow from the north. These reversals occur because low pressure created by hot air rising from summer-heated continental land-masses draws in wind from the opposite, cooler hemisphere. It is a marvellous and unique atmospheric engine which, as has been pointed out earlier, produced the climate to provide good rainy conditions for evolutionary progress in East Africa (and hence the Nile valley) and in Mesopotamia, India and south-east Asia. In addition, this engine moved ships back and forth across the northern Indian Ocean and up and down it as far as the tropics. Lightly-constructed vessels with square sails could use these reciprocating winds since they are quite accurately predictable.

Areas of potential seasonal storms could be defined and understood, and therefore usually avoided.

There is no doubt that seatraders had the ability and knowledge to reach the East African coast long before the first millennium BC. The problem has always been the question of an incentive and lack of information about who they might have been trading with. Egyptians, Indians, Israelites, Phoenicians, Nabateans, southern Arabians and Persian Gulf people were seamen and merchants. They did not set up colonies to do their own hunting for ivory, metal mining and the gathering of valuable goods for seaborne trade. They needed local people with whom to build a trading tradition. There were suitable peoples, proved by history, classical literature and archaeology, along the African Red Sea coast and around the Horn, and the hinterlands, but what about East Africa?

The first millennium BC saw gradually increasing disorder in the whole region of ancient civilisation as populations grew and the climate changed; fertile lands becoming increasingly inhospitable under the onslaught of intensive agriculture and declining rainfall. Trees were cut out for building materials, ships and firewood. Metallurgy and large urban populations need great masses of firewood and the reason for the decline of the great Mesopotamian and other regional cities can easily be proposed to be the exhaustion of tree stocks on the fringes of expanding deserts. Destruction of forests and scrub bush accelerated desertification.

Recently, a strange piece of intelligence was provided by Dr Shared Master of the University of the Witwatersrand in South Africa. Robert Mathews in *The Sunday Telegraph* of 4th November 2001 described how Master and colleagues had analysed satellite images which showed a peculiar and regular circular feature in the Al 'Amarah region near the confluence of the Tigris and Euphrates Rivers in Iraq. Draining of the marshes by Saddam Hussein's regime to combat the Marsh Arabs in the 1990s had revealed this crater-like geological shape. Other evidence suggested that a meteor struck this place about 4,300 years ago, and it would have contributed to the sudden decline of Mesopotamian civilisations. The decline in Persian Gulf activities, apart from general political changes of the whole power structure of the whole geographical zone, would have opened new niches of opportunity for Red Sea seatraders.

Nation states thrashed about. Israel was ravaged by Egyptians and Egypt itself began a slow decline, invaded by Assyrians, ruled by Libyan and Nubian pharaohs. During the 26th Dynasty (663-525 BC) there was a brief resurgence of power through sea trading in the Mediterranean and the Indian Ocean with the help of Phoenicians.

King Neccho II, a Libyan of that 26th Dynasty, hired a Phoenician fleet to circumnavigate Africa sometime shortly after 600 BC; the only recorded voyage between the Indian and Atlantic Oceans until the Portuguese achieved it again 2,000 years later. There is controversy about the motive for this exploratory voyage and this has fuelled doubts as to its authenticity: why should Neccho have commissioned such an extravagant expedition? It suggests to me that there was firm belief in the concept of an ocean route around Africa and that some knowledge of the coasts to the southwards on both sides of the continent was already available. Problems with navigation in the Mediterranean safe from rival nations is the usual explanation for this enterprise, and it is a reasonable one. The Portuguese drive to reach the Indian Ocean around the Cape was for the same reason. Maybe it was the Phoenicians, cut off from the western Mediterranean by the independence of their own colony of Carthage, who had the urge to explore and petitioned the Pharaoh who was sympathetic and ready to try new markets. Neccho was keen on trading and had the ancient canal linking the Mediterranean and the Red Seas dug out after it had silted from centuries of disuse.

Amongst many stories of seafaring, Herodotus describes the Phoenicians' voyage around Africa and a subsequent attempt by a young Persian prince who was tasked to do the same as a penalty for violating a woman. These voyages have often been disputed by historians, but I have no doubt that they occurred and I make no apology for attaching Herodotus' text here (as translated by Macauley, by courtesy of Gutenberg Books).

> I wonder then at those who have parted off and divided the world into Libya [North Africa], Asia, and Europe, since the difference between these is not small; for in length Europe extends along by both, while in breadth it is clear to me that it is beyond comparison larger; for Libya furnishes proofs about itself that it is surrounded by sea, except so much of it as borders upon Asia; and this

fact was shown by Necos king of the Egyptians first of all those about whom we have knowledge. When he had ceased digging the channel which goes through from the Nile to the Arabian gulf, sent Phoenicians with ships, bidding them sail and come back through the Pillars of Heracles to the Northern Sea and so to Egypt. The Phoenicians therefore set forth from the Erythraean Sea [Indian Ocean] and sailed through the Southern Sea; and when autumn came, they would put to shore and sow the land, wherever in Libya they might happen to be as they sailed, and then they waited for the harvest: and having reaped the corn they would sail on, so that after two years had elapsed, in the third year they turned through the Pillars of Heracles and arrived again in Egypt. And they reported a thing which I cannot believe, but another man may, namely that in sailing round Libya they had the sun on their right hand.

Thus was this country first known to be what it is, and after this it is the Carthaginians who make report of it; for as to Sataspes the son of Teaspis the Achaimenid, he did not sail round Libya, though he was sent for this very purpose, but was struck with fear by the length of the voyage and the desolate nature of the land, and so returned back and did not accomplish the task which his mother laid upon him. For this man had outraged a daughter of Zopyros the son of Megabyzos, a virgin; and then when he was about to be impaled by order of king Xerxes [who reigned 485-465 BC] for this offence, the mother of Sataspes, who was a sister of Dareios, entreated for his life, saying that she would herself lay upon him a greater penalty than Xerxes; for he should be compelled (she said) to sail round Libya, until in sailing round it he came to the Arabian gulf. So then Xerxes having agreed upon these terms, Sataspes went to Egypt, and obtaining a ship and sailors from the Egyptians, he sailed to the Pillars of Heracles; and having sailed through them and turned the point of Libya which is called the promontory of Soloeis, he sailed on towards the South. Then after he had passed over much sea in many months, as there was needed ever more and more voyaging, he turned about and sailed back again to Egypt: and having come from thence into the presence of king Xerxes, he reported saying that at the furthest point which he reached he was sailing by dwarfish people, who used clothing made from the palm-tree, and who, whenever they came to land with their ship, left their towns and fled away to the mountains: and they, he said, did no injury when they entered into the towns, but took food from them only. And the cause, he said, why he had not completely sailed round Libya was that the ship could not advance any further but stuck fast. Xerxes however did not believe that he was speaking the truth, and since he had not performed the appointed task, he impaled him, inflicting upon him the penalty pronounced before. A eunuch belonging to this Sataspes ran away to Samos as soon as he heard that his master was dead, carrying with him large sums of money; and of this a man of Samos took possession, whose name I know, but I purposely pass it over without mention.

The reference to the sun being on the right hand, which Herodotus finds unbelievable, is shown to be superficial proof of the accuracy of the report. Herodotus and others of his time would not have easily understood that south of the Tropic of Capricorn, no matter what the season, the sun always lies to the north (on the right hand) as the voyagers traversed the southern coast of Africa.

The punishment voyage of the Persian prince is not often quoted, but it also has a great ring of truth. Clearly, he reached the West African coast south of the Sahara where he met Late Stone Age forest-dwellers.

The Carthaginian, Hanno, followed in the wake of the Persian prince about 450 BC. He was charged with setting up trading colonies and set off with an armada and large numbers of people, dropping some off at suitable places along the coast. There is no doubt that his expedition occurred, but there is controversy about how far he reached. Lionel Casson in *The Ancient Mariners* (1959) presents a measured opinion:

> Just how far did Hanno get? Most geographers hold that he stopped short of the calms and heat of the Gulf of Guinea and pushed no further of Sierra Leone. ... Others take him as far as the Cameroons, arguing that the Chariot of the Gods [the terminal landmark described by Hanno as a mountain with a leaping flame at the centre] is better identified with Mount Cameroon, the tallest peak in West Africa and a volcano to boot.

One cannot forget the colonisation of Pacific Islands by navigators of south-east Asia and Indonesia. However, like the modern Arab and Indian oceanic sailing vessels, Indonesian and Polynesian craft were designed and used in the tropical oceans by navigators who were masters of the climate cycles and made use of the tropical wind systems. There can be no doubt that the regions outside of the trade wind and monsoon belt were explored by these early oceanic navigators, but I am also sure that having discovered that the frequently stormy weather was quite unsuitable for their vessels, and that the temperate lands were more-or-less alien to them, subtropical regions bounded by the Southern Ocean were shunned. It is notable that the South Island of New Zealand was possibly not colonised by Polynesians until the 12th or 13th century AD and that event was a migration southwards from an earlier colonisation of the North Island, not directly across the ocean.

I have no doubt that West Africans were oceanic voyagers in large dugout canoes, similar to those used by Polynesians. I have seen the great canoes still in use along the Ghanaian coast by offshore fishermen. My correspondent Harry Bourne has followed traces of evidence that West Africans traversed the tropical belt to the Caribbean and Central America in pre-Colombian centuries and I can have no dispute with the probability. The winds and currents ensure safe voyaging to and fro and West Africans were certainly capable of celestial navigation. They had an ancestry partially descended from Late Stone Age people with long experience of Saharan nomadism and celestial navigation in some form was undoubtedly known and used by Saharan peoples from at least 5,000 years ago. This aspect of African history and culture is fascinating, but is beyond the immediate scope of this book which is primarily concerned with the Indian Ocean.

Andrew Collins, who has written extensively about megalithic civilisations, pointed out to me that Hanno's celebrated voyage has to be just one important oceanic expedition by Carthaginians. We are aware of it through Herodotus' brief description, and it is clear that these great seatraders made many voyages in the Atlantic. In any case, it is obvious that no ruler or his government would have sanctioned the cost and use of much manpower for the planning and outfitting of a large colonial expedition unless there had been precedent and a fair belief in success. The Carthaginians were traders and their heritage was maritime. Collins reminded me of the intense rivalry between the Mediterranean maritime trading nations and the vicious and prolonged warfare between Greeks and Phoenicians, and then Greeks and Carthaginians after the decline of Phoenicia. When Rome entered the arena it was a fight to the death between the three powers with the Romans conquering all in the end. This fierce rivalry provoked great secrecy, especially by the Carthaginians, and it has to be remembered that all of Carthaginian literature was destroyed by the Roman conquerors. The only knowledge we have of their activity is drawn from Greek and Roman sources.

There is little doubt that Carthaginians, building on the knowledge of their Phoenician origins, explored extensively in the Atlantic. They knew the Sahara trade and sent explorers overland to West Africa. Their subject people, the Berbers, had been established as Neolithic agriculturalists for millennia in contact with Egypt. There was no lack of knowledge of that region of Africa, the western coast of Europe encompassing the British Isles, and the central Atlantic. The Hanno expedition was founded on knowledge and was prompted by the need to replace the metals trade with southwest Europe after this was interrupted during one of the many trading wars.

Professor Felix Chami believes that Carthaginians persisted in their Atlantic colonising efforts and founded a colony in southwest Africa, seeking the metals of southern Africa. As mentioned in a previous chapter, he has proposed that it was they who introduced large numbers of Berber soldiers and workers who brought with them their genes and culture which contributed to the founding of the Khoi people of the Cape. I see a number of serious difficulties in this particular far-fetched speculation of Carthaginian maritime trading and colonial activity, but that does not dim the shining importance of oceanic seatrading to tropical Africa.

More conventionally, Felix Chami has referred to the spice trade between the Indian Ocean and Europe in several publications, quoting classical sources. There seems no doubt that oriental spices were brought to East Africa by sea, perhaps in Indian and Indonesian vessels and then transhipped overland to Mediterranean ports and thence to Rome. Spices were always very high-value merchandise and it is reasonable that the fastest way to get easily transportable goods from Indonesia to Rome was this method, eliminating the waiting about in East African ports for other cargo, refurbishing and the changing of the monsoons. Indonesians may have initiated this trade through their own agents on the mainland in order to steal a march on the Egyptian merchants whose route was much slower.

Roger Blench in his preliminary paper presented at the conference on the Maritime Heritage of the Indian Ocean in Zanzibar in July 2006 proposes two distinct colonisations of the western Indian Ocean by Austronesians. The first occurred during the first millennium BC when I surmise that exotic foods were introduced, and the second is identified in 5th - 7th centuries AD. It may be assumed that the earlier, least identified migration touched at both the shores of Madagascar and the African mainland. Linguistic evidence helps to clarify the later colonial migrations which began the thorough population of Madagascar. The modern Malagasy language has been linked to the Barito languages of Borneo, and the Sama-Bajaw 'sea-people' who roamed the Sunda seas are part of that group. However, there are Malay links which Blench suggests may be the reason for the migration at that time because there were imperial Malay activities then which could have precipitated turmoil.

In a summary of his paper, Roger Blench states:

> ... Previous models have tried to evolve a single model to account both for the settlement of Madagascar, and contacts with the coast despite the chronological problems. It is here proposed that the two migrations were essentially unrelated and that the Austronesian navigators who were crossing the Indian Ocean at the earlier date, probably did not come from Borneo, but a different region of insular SE Asia, perhaps the Philippines.

Whether the Malagasy people are descended from Austronesians who directly colonised Madagascar, or who first attempted settlement on the mainland and were made to be unwelcome by coastal Bantu-speakers has been under discussion. The Malagasy are a complex people with different tribal or community groups with distinct traditions, and the history of Madagascar is a fascinating microcosm of modern human settlement of a large uninhabited land. There is historical evidence that Africans were taken to the island as slaves or clients over long time, and there is no reason why Africans should not have migrated for their own advantage at various times. Cattle and other domestic animals were certainly introduced from the mainland. Madagascar was used by European and American pirates as a base during the early period of European expansion into the Indian Ocean, and there is substantial historical evidence of Malagasy pirates and slavers ravaging the Comores and further afield. Piracy as well as more conventional exchange of goods through trade has long been a custom in Indonesia.

Oceanic voyaging in large canoes or plank-built vessels, propelled by sails, was far more prevalent for many centuries throughout the tropics than is often acknowledged, especially from a Eurocentric viewpoint. The use of celestial navigation and its practice in the Pacific and Indian Oceans has been published, most notably by G.R.Tibbetts in *Arab Navigation in the Indian Ocean before the coming of the Portuguese* (1971) and by David Lewis in *We, the Navigators* (1972).

As I have discussed elsewhere, my most persistent thinking about early oceanic voyaging and the founding of pre-Christian era colonies on the African coasts is always shadowed by consideration of the geography. All the ships designed for use in the northern and tropical Indian Ocean were intended to sail in the favourable monsoon. There are hurricanes and other severe tropical storms in the Indian Ocean, but their seasons were understood and navigators would avoid them. The entrance to the Red Sea and the Gulf of Aden experience regular gale force katabatic winds especially during the hottest summer months, but they moderate away from the coast and are well-known to local sailors.

The west and east tropical coasts were congenial to seatraders and colonists, but it is significant that so far there is only archaeological and historical records of ancient voyaging to the East African coast. Prof. Felix Chami's reports of archaeological results from sites on Kilwa Island, Mafia Island, Zanzibar and the Tanzanian mainland show convincing evidence of oceanic contact from at least 3,000 years ago and this was certainly regular activity toward the end of first millennium BC. Apart from Hanno's colonising voyage, and other possible attempts which are unknown to us, there is no evidence of northern hemisphere people having regular contact with the Guinea coastlands until the Portuguese in the 15th century. That West Africans have a long tradition of seagoing in their giant canoes is fact and they may have voyaged to the Caribbean, maybe inspired by diffusion of knowledge from Carthaginians or others, but it seems that it was learned that economical trading was better pursued with trans-Sahara caravans, even when it was only practicable with camels.

South of 25ºS. latitude the situation changes dramatically on both sides of the continent. Gales are frequent and unpredictable and there are strong ocean currents. The Cape of Good Hope has been notorious for sailors since the first Portuguese passed it in 1488. Many large and heavily constructed European sailing vessels have been wrecked or lost from the 16th century onwards and modern ships have been broken at sea or stranded. On the west and southern coasts the land is harsh and uninviting with cold winters, and no Europeans attempted to found a colony until the Dutch in 1652 when obliged to do so because of continuing and increasing loss of men and ships from malnutrition, fatigue and the need for repairs.

The people living on the inhospitable southern coast of South Africa have been explored thoroughly in the last thirty years by archaeology at a number of sites with dates far back into the Middle Stone Age. Continuity has been established during the changes in culture and technology experienced by the ancestral San from before 100,000 years ago to the arrival of domestic animals at about the beginning of the Christian era. Early Iron Age agriculturalists, presumed to be Negroid and Bantu-speaking, never penetrated the south-western Cape and there has not been any archeological trace of ocean voyagers before the Portuguese.

*   *

Carved in stone on monumental buildings is the first recorded voyage down the Red Sea. In 2750 BC, the Egyptian explorer, Hannu, sailed to Punt which is known to have been the lands about the Horn, or even further south along the coast of East Africa. Where Hannu explored, there can be no doubt there were many hundreds and probably thousands of unrecorded routine trading voyages in the centuries which followed.

Similarly with navigators from the Persian Gulf. There is a persistent clue to Persian Gulf contact with East Africa from the first millennium BC. It is the *siwa* horn, a ceremonial instrument, long and curved in the symbolic shape of an elephant's tusk which was blown to draw attention to a ruler's presence or some official occasion. Often, it was used as a symbol of authority like a mace or ceremonial sword. The *siwa* horn survived in East Africa throughout the Islamic Arab period and tourists can buy replicas of them in Lamu or Mombasa curio shops today. Some scholars have identified it directly with Assyrian horns of 3,000 years ago, used for the same purpose. Others have dismissed the connection, not because it was impossible, but because of an opinion that such early contact was unlikely.

However, it has to be remembered that Assyrians became bold sailors when their empire encompassed Babylonia; they conquered Egypt and ruled the Red Sea trade. Sea trade maintained its momentum and I believe it was during this time, the middle centuries of the first millennium BC, that trade with East Africa became firmly established and the siwa horn is a thin thread of evidence.

Whatever happened to the fortunes of Egypt, Arabia, Ethiopia, Mesopotamia and India, as dynasties rose and fell and one kingdom or empire succeeded another, the Indian Ocean trading system prevailed. The monsoons were too convenient, the variety of goods within the easy limits of navigation too tempting and the energy of ordinary merchants and sailors sufficient to keep it going whatever political events intervened. Along the northwestern rim of the ocean, city-states rose when older ones failed. Trade may have faded and revived, but it never stopped.

After Alexander the Great conquered all these lands and moved onto the Indus plains of northern India, Greek influence and the activity of Greek merchants spread far. During wars in the eastern Mediterranean, naval fleets became strategic weapons as potent as a modern nuclear arsenal. Greeks and Trojans learned early and the Hittites and Philistines taught Egypt the lesson. For example, Alexander's admiral, Nearkhos, transported an army from Thatta in the Indus delta, in today's Pakistan, to the Persian Gulf by sea. Phoenicia created the first maritime empire.

A Greek, named Diogenes, is the first identified foreign person to have seen a snow-capped equatorial mountain in Africa, most probably Mount Kilimanjaro, having travelled up the Pangani River. That started the remarkably accurate legend of the mystical African 'Mountains of the Moon' which sourced the Nile and were described by Ptolemy. This legend was still inspiring Livingstone and other explorers in the 19th century.

When Roman Civilisation began running with accelerating speed, its appetite for luxurious and exotic goods became voracious. The Roman Empire was arguably the greatest trading machine the world had produced. Incense, crystalline rocks and stones, refined metals and artifacts, ivory, exotic animals and their products, fine cloth, rare woods, all were increasingly desired around the Mediterranean and the interior of western Europe. Roman fleets sailed the Indian Ocean whether passing through the Nile delta by canal or built in Egypt. Strabo recorded that an annual Roman fleet of 120 vessels sailed to India at about the time of Christ.

Herodotus and Pliny compiled important collections of historical information. Ptolemy is the best-known of early geographers, basing his writings on the lost books of Marinos of Tyre, who recorded trading voyages to East Africa, and compiled interpretations of information from many other sources for his famous *Geographia*. There are Indian epics which suggest that their explorers penetrated as far as the Great Rift Valley lakes. Persians and Arabs had their stories of Indian Ocean voyaging.

The great mass of ancient writings which were lost in the sacking of cities such as Jerusalem and Alexandria and the destruction of their libraries were available then within the sweep of literate society from the Mediterranean to India. Much scholarship which flourished and retained knowledge in the great Islamic universities in Spain was lost during the wars to expel the Moors. The destruction of the library at Constantinople where much was preserved was a great disaster in the 15th century. The important surviving works of Herodotus, Pliny, Strabo and Claudius Ptolemy were their own selections from and interpretations of other reports and books which have been lost. The same applies to the maps based on earlier works which were produced in the 16th century. Our knowledge today of what was going on is only a few outline sketches of sections of a great canvas.

SIX : *THE SWAHILI COAST*
The impact of seatraders on eastern Africa.

It is the *Periplus of the Erythraean Sea* which is the earliest clear and detailed historical record of sea trading with East Africa which has been handed down to us. Its date has been bandied about by scholars but is now accepted to have been written about 100 AD.

It was a guide, or practical manual, to the ports and markets of the Indian Ocean written by an anonymous Alexandrian Greek, briefly describing the rough distances usually in terms of easy sailing stages, the landfalls, towns and their inhabitants, the regional economy and the trade that could be expected. It is an extraordinarily valuable document for anybody interested in eastern Africa, Arabia and India and their ports on the Indian Ocean known to the author and his associates and contacts at that time. Its simplicity and lack of extravagant language or editing gives it potency.

G.W.B. Huntingford's translation and commentary, published by The Hakluyt Society in 1976, was the standard reference for some years and remains a valuable source. Lionel Casson's version, published in 1989 by Princeton University Press, is now the more important reference because it had up-to-date background from later archaeology and literary research. Unfortunately, at the time of writing, these books were out of print and difficult to get hold of. I hope the publishers will remedy this.

The *Periplus* describes the Red Sea coasts, East Africa as far as Zanzibar and the Tanzanian coast, southern Arabia and around India from the Indus to the Ganges. The unknown author did not know much about the Persian Gulf; perhaps he never sailed there, probably Arabs and Persians of the time did not encourage Egyptian and Greek traders, requiring them to transship their cargoes at an entrepôt such as Salalah, or Taqa in the kingdom of Dhofar on the southern coast of Arabia.

Most of the seaborne African trade at that time, according to the *Periplus*, was with the markets on the coast of modern Sudan, Eritrea and the Horn of Africa. These places were within easy caravan distance of the cities of the southern Sudan, principally Meroe (near modern Khartoum), and Axum in northern Ethiopia, the capital of an impressive empire for a thousand years. There was a rich range of products traded with these wealthy and sophisticated northeast African kingdoms. Huntingford's edition provides this information which I have summarised:

> IMPORTS FROM EGYPT - Unscoured Egyptian cloth, Arsinotic robes, spurious coloured cloaks, fringed mantles, linen cloth, several sorts of glassware, imitation murrine ware (a semi-precious crystal) made in Diospolis, rokhalkos used for ornaments and money, material called 'copper cooked in honey' for pots and armlets and anklets, iron used for spears for hunting elephants and for war; axes, adzes, swords, big bronze drinking-cups, Ladikean and Italian wine.
>
> For royal customers - Silver and gold objects custom-made in the design of the country, cloaks of cloth, unlined garments not of much value (presumably for servants, soldiers and slaves).
> IMPORTS FROM ARIAK (the Gulf of Cambay in India) - Indian iron and steel, broad Indian cloth, cloth called Sagmatognai, belts, garments called Gaunakai, mallow-cloth, muslin cloth and coloured lac (a dark-red resin).
> EXPORTS FROM THE AFRICAN RED SEA PORTS - Ivory, Tortoiseshell and Rhinoceros horn.

One is reminded of a catalogue that a 19th century Liverpool merchant might have offered to a trading house in Zanzibar 1,800 years later. The kingdoms of Meroe and Axum must have exported much ivory in exchange for all those manufactured goods from the industrial nations of the day. It is not surprising, therefore, that elephants became rare in those parts and ivory trading had to move on to East Africa.

Beyond the narrow strait of Bab el Mandeb dividing Africa from the tip of Arabia at the Horn of Africa, spices, incense (named 'from beyond the straights'), special woods, frankincense and a variety of other aromatic gums were exported, some of it in local vessels across to the Yemen.

Huntingford translates the description of Mouza, a major trading port in Yemen :

> The whole place is full of Arabs, shipmasters and sailors, and hums with business: for they use their own ships for commerce with the opposite coast [of Africa] and with Barugaza [in the Gulf of Cambay, India].

This southern Somali coast was called Azania, (derived from "dried" or "parched" and used by Pliny and Ptolemy as well as in the *Periplus*). Azania was somewhat barren until the East African markets were reached: the Puralaon Islands (Lamu Archipelago), Menouthias (Pemba Island or Zanzibar) and Rhapta (a 'lost' town identified with somewhere on the Tanzanian coast, *see below*). On Menouthias the locals were fishermen, working with sewn boats and basket-ware fishtraps. Sailors and travelling merchants knew the people of Rhapta well. The trade was important but not as sophisticated as that with the Red Sea coast and the Horn. The Huntingford edition defines the trade at Rhapta:

> IMPORTS: Spears, axes, small swords, awls, several kinds of glassware, wine and corn (all manufactured in Mouza on the Red Sea shore of today's Yemen).
> EXPORTS: *Much* ivory, rhinoceros horn, tortoiseshell and a little coconut.

It can be seen from these brief catalogues that it was a complex business. A merchant sailing in his own tramping vessel along that coast had to have an intimate knowledge of local needs and customary rules the equal of any equivalent in the modern European era. His product knowledge and financial planning had to be excellent for sustained success. When one adds the hazards of sailing ships with sewn planking without modern navigation aids to the commercial problems, then one realises that the Indian Ocean seatraders 2,000 years ago were intelligent and experienced, brave and bold entrepreneurs.

It must also be recognised that the system that was working at that time could not have arisen suddenly. It was developed over time and therefore it must also be recognised that if Rhapta was well-established by 100 AD, the eastern African coast as far south as

Mozambique had not only been explored geographically, but organised for trade before the beginning of the Christian era.

It seems obvious that the seatraders of the time of the *Periplus* were in contact with elements of Late Stone Age agriculturalists from the north or west who had settled the hinterland of Rhapta. It is possible that some may have been nomadic herders converted to cultivation and fishing for subsistence wherever cattle were inapplicable because of disease. Richard Wilding in his *Shorefolk* (1987), and in conversation with me, suggested Cushitic, and that conclusion seemed reasonable to me at the time. But latest pottery evidence presented by Chami and others shows that they had Nilotic traditions with far contact into the Nile region of the Sudan.

These people were amenable to trading with strangers and became clients to seatraders from Arabia. Since they traded imported iron tools and weapons from Arabia, it may be concluded that they were apparently not Bantu-speaking Early Iron Age farmers from the interior who settled later at places such as Kwale near Mombasa.

Mindful of the influence that trading has on people generally, I firmly speculate that there was a symbiotic beginning to both extended settlement of the coast and the commencement of trading. People moved down the coast pursuing agricultural advantage in a time of changing population and climate, and coincidentally they were people with whom seatraders could easily encourage a trading connection, with the necessary gathering of desired raw materials and the distribution of imported manufactured goods. Research is being pursued which may show a regular connection over several centuries in the first millennium BC between the Nile, the East African highlands and the Lacustrine Zone generally, and down to the coast via river roads. These are routes for people and their hominid ancestors which have been used since the dawn of humanity.

In 1989 the new translation of the *Periplus* was published by Lionel Casson, titled *The Periplus Maris Erythraei*, with his notes and commentary.

There is a particularly important key word which Casson re-translates and interprets. Huntingford describes the people of Rhapta (Chapter 16) :

> From here after two courses off the mainland lies the last mart of Azania, called Rhapta, which has its name from the aforementioned sewn boats, where there is a great deal of ivory and tortoiseshell. The natives of this country have very large bodies and piratical habits ; and each place likewise has its own chief. The Morpharitic chief rules it [Rhapta] according to an ancient agreement by which it falls under the kingdom which has become first in Arabia. Under the king the people of Mouza hold it by payment of tribute, and send ships with captains and agents who are mostly Arabs, and are familiar through residence and intermarriage with the nature of the places and their language.

William H Schoff''s earlier translation (published in 1912) reads:

> 16. Two days' sail beyond, there lies the very last market-town of the continent of Azania, which is called Rhapta; which has its name from the sewed boats (*rhapton ploiarion*) already mentioned; in which there is ivory in great quantity, and tortoise-shell. Along this coast live men of piratical habits, very great in stature, and under separate chiefs for each place. The Mapharitic chief governs it under some ancient right that subjects it to the sovereignty of the state that is become first in Arabia. And the people of Muza now hold it under his authority, and send thither many large ships, using Arab captains and agents, who are familiar with the natives and intermarry with them, and who know the whole coast and understand the language.

Casson's translation of the same chapter concerning Rhapta :

> Two runs beyond this island [Menouthias, probably Pemba or Zanzibar] comes the very last port of trade on the coast of Azania, called Rhapta, a name derived from the aforementioned sewn boats, where there are great quantities of ivory and tortoise shell. Very big-bodied men, tillers of the soil, inhabit the region; these behave, each in his own place, just like chiefs. The region is under the rule of Mapharitis, since by some ancient right it is subject to the kingdom of Arabia as first constituted. The merchants of Muza hold it through a grant from the king and collect taxes from it. They send out to it merchant craft

> that they staff mostly with Arab skippers and agents who, through continual intercourse and intermarriage, are familiar with the area and its language.

It should be noted that, firstly, Rhapta is described by the three translations as having an ancient subordinate relationship to Mouza / Muza in the Yemen and that Arabs who have intermarried with locals are favoured as captains or supercargoes. This indicates that Rhapta has a long-standing position as a trading town with some kind of monopoly claimed by Sabaean seatraders. The key to the people of Rhapta in which the three translations differ are the two meanings of a Greek word which could be interpreted as either describing pirates or farmers, with different nuances. It is not easy. Huntingford spends much time over the word in his notes and decides to use 'piratical habits' because when he was writing, in the 1970s, it was not considered that agriculture had spread to the Tanzanian coast. Casson, writing maybe fifteen years later, had the benefit of new archaeological discovery and academic opinion. 'Tillers of the soil' is clearly the correct interpretation and of great significance. Here is the first historical reference, presumably firsthand, to agriculture on the East African coast.

Professor Felix Chami of the University of Dar es Salaam has been working for many years on the archaeology of the Tanzanian coast and, like a number of us, has obviously been captivated by the challenge of unravelling the mysteries of early contact and trading links between the Indian Ocean coast of East Africa and the kingdoms of Arabia, Egypt and the Mediterranean.

I first heard Chami deliver a paper in Cambridge, England, in 1994 when he described his excavations at that time and his interpretation which showed a continuity and growth of people, and their trading activity from the 1st to 8th centuries on the Tanzanian coast, characterised by what Chami calls TIW (triangular incised ware) pottery and which others have traditionally called Tana-River pottery. Chami was exploring with archaeology and commentary an era and area often missed by other professionals or scholars, and usually considered to be dominated by the Early Iron Age of Bantu-speakers with Kwale style pottery. (Chami's paper: *The first millennium AD on the East Coast, a new look at the cultural sequence and interaction.* Azania XXIX-XXX, 1996).

Dr Mark Horton, an archaeologist who worked for some years earlier on Swahili-Arab sites in East Africa, had suggested that the spread of Tana-tradition pottery during the period covered by the *Periplus* indicates that the inhabitants of Rhapta were descended from a 'Pastoral Neolithic' people of probable Cushitic-speaking cultural origin. This view had also been part of Richard Wilding's thesis in his *Shorefolk* (1987). Horton described how their pottery, known as Tana River pottery different from Early Iron Age Bantu pottery of the Kwale style, had been found all along the coastal region of Kenya and Tanzania with a variety of dates from the late 1st millennium BC onwards. It had been tentatively recognised at Chibuene in southern Mozambique, excavated by Paul Sinclair, and associated with Sofala, together with Kwale-type. I interpreted this as indicating that a complex of Negroid people of both Late Stone Age and Early Iron Age industries, with a mix of Nilotic, Cushitic and Bantu cultures were settled all along the eastern African shores in the first millennium AD.

Felix Chami's work progressed and he published on several occasions. I find his particular papers, *Neolithic Pottery Traditions from the Islands, the Coast and the Interior of East Africa* (2003) and *Kaole and the Swahili World* in the Dar es Salaam University publication, *Southern Africa and the Swahili World* (2002) to be significant. His principal archaeology has been centred on Mafia Island and offshore associated islets and the mainland around the Rufiji River delta and the coast north of Dar es Salaam as far as Bagamoyo. Obviously Zanzibar and Pemba Islands are within his area of interest and he has more lately published on his finds on Zanzibar.

Ptolemy suggested three clues to the position of Rhapta : it was at the mouth of a river, the town itself was some way back from the coast or port, and it was at a cape. Ptolemy was quoting other travellers, Diogenes (the first to confirm snow-topped peaks on the Equator) and Marinos of Tyre, both of whom are assumed to have actually visited Rhapta. When Huntingford published his *Periplus*, five possible locations for Rhapta were proposed by scholars attempting to equate modern places with the descriptions in the *Periplus* itself and Ptolemy's geography. These were Pangani, Tanga, Msasani near to Dar es Salaam, Kisiji and the Rufiji Delta. General opinions seemed to favour the Rufiji.

Chami's excavations in the 1990s show that the Rufiji Delta area and the immediate offshore islands of the Mafia group can almost absolutely be identified with Rhapta. Late Stone Age pottery of Chami's TIW (Tana River style) have been discovered with dates to the latter half of the first millennium BC, which together with other evidence proves the presence of Neolithic agriculture and seatrading. Two of Ptolemy's criteria were immediately satisfied: the presence of a river, and the Rufiji is one of East Africa's most important, and that the town was back from the port. Chami's excavations show that a main centre was at Mkukutu-Kibiti (about 35 kms from the sea) with seatrading evidence from Roman beads. The placing of a 'metropolis' or 'capital town' of the Rhapta people, as described by Ptolemy, away from the sea could have both strategic and geographical reasons. It was away from the immediate danger of any oceanic rivalry and away from the heaviest rainfall of the coastal monsoon.

There is no prominent cape on that low-lying coast, but the Rufiji delta protrudes some distance from the mostly regular shoreline. The first, most northerly evidence of it could be seen as a landmark cape by approaching sailors. Mafia Island lies immediately offshore and could be seen as a landmark from a distance.

Another possibility has been tentatively suggested and for a time I was strongly attached to it, and that is the medieval trading centre of Kilwa. Kilwa became the wealthiest Arab-Swahili trading city in eastern Africa during the height of the gold trade with Zimbabwe in the 14th-15th centuries AD and I have often believed that places which are chosen and flourish under one people must be equally attractive to others. I visited Kilwa and the Rufiji area on a personal quest to see the famous medieval ruins and my romantic attachment to the area was established. But, of course, the reasons for founding a trading city on an island with a good harbour by Arab seatraders in the second millennium AD would be very different from the needs of Neolithic farmers of the first millennium BC. The people of Rhapta, I had to remind myself, were primarily cultivators and some were fishermen. Foreign trade was a brief and periodic luxury. Ships would have come once a year, and often there may have been gaps of several years. However, my intuition regarding Kilwa eventually proved to be well-founded and Chami's later publication *The Unity of Ancient African History 3000 BC to AD 500* (2006) states:

> Other more recent findings and observation include: a current research on the island of Kilwa, renowned for its Triangular Incised Ware tradition (Chami 1994) or kitchen ware (Chittick 1994) and as the trading centre for the Swahili coast between AD 1000 and 1500, has uncovered many pottery of Neolithic tradition, suggesting a sequence running through Kansyore / Nile Valley Neolithic, Nderit and Narasura [Great Rift Valley]. This pottery has been found on the surface contexts mixed with stone tools.

Chami's conclusions are wide and comprehensive. He has gone further than other East African research and I find his work in this context to be fascinating in its illumination of the pre-Christian era in the western Indian Ocean. He goes on :

> These incontrovertible findings truly suggest that there was no pottery tradition in the African Neolithic [Late Stone Age] period specifically belonging to one particular region of eastern Africa or another. The people of the Nile, Great Lakes and the Rift Valley regions, not isolating the rest of the Sahel, were culturally linked. ... These cultural links were not necessarily reflecting people's migration, but some kind of cultural fluidity though contact.

There may be dissenters from this broad statement, but it resonates with me. It was the exploitation of Iron, from which followed the burgeoning of cultivation and population explosion in the Interlacustrine Zone, coincident to the dessication of the Sahara, which caused migrations and the arrival of new people on the coast, adding to the Swahili mix. Increasing trade along the coast precipitate more change and the surging movement of people with Early Iron Age culture southwards in the first millennium AD.

The pattern of contact with East Africa that had been established in the first millennium BC continued after the time of the *Periplus of the Erythraean Sea*. It waxed and waned, but there were merchants and sailors who knew where to go and what rewards there were. The engine of the Roman Empire that generated trade wherever ancient civilisation had placed its mark had important effects on eastern Africa which persisted into the 4th century AD. J.E.G.Sutton in the *General History of Africa* (1981) wrote:

> The demand for ivory grew enormously as the Romans began to use it not only for statues and combs but also for chairs, tables, bird-cages and carriages; there was even an ivory stable for the imperial horse.

Ivory increasingly had to come from Africa. Until the metallurgical exploration of southern Africa had been completed by Early Iron Age people seeking metals, which resulted in the discovery of gold along the Limpopo River and Zimbabwe, ivory was the driving force of regional trade.

In my opinion, the ivory trade was the trigger firing the cannon of population growth which precipitated Bantu-speaking colonisation of southern Africa.

* *

Indonesians sailed across the ocean and colonised Madagascar and some may have settled on the adjacent mainland coast as farmers but also as traders. There is the probability that they initiated a fast-track spice trade overland to the Mediterranean from East African ports. In medieval times, Indonesians were notorious pirates in the region of the Comores. Indians from Cambay and the trading city-states along the Malabar coast were active. Roman traders, whether Italians or natives of other parts of the Empire, roamed widely. Roman coins and beads have been recovered along the East African coast. Madurai, a trading city on the eastern side of India, was a particular outpost of Roman trade. St.Thomas, Jesus' apostle, went to India and began a Christian tradition taken up by missionaries from Syria and Persia that persists until today. Indian sailors were active on the East African coast. Apart from the direct trading voyages of the great Chinese admiral, Zheng He, who commanded several expeditions with fleets of huge junks into the western Indian Ocean in the 15th century AD, India was always a transhipment centre for goods travelling between Europe, Arabia, Africa and the Far East. Chinese traditions may be seen persisting on the Malabar Coast until today.

2005 was the 600th anniversary of the first of seven voyages of exploration and trading that Zheng He commanded in the Indian Ocean. After the first explorations, determining the geographical and political structures, the fleets would break up to serve particular markets. Malacca in Malaysia and Calicut and Cochin on the Malabar Coast were their bases and it is in Cochin, 600 years later, that the evidence of Chinese presence is most obviously seen. Unsurprisingly, the Portuguese, following the Chinese, found these three eastern entrepôts to be the most important. Zheng He's first fleet was of sixty-two large vessels of about 400 feet in length and nine masts, accompanied by up to 250 smaller junks.

> [The fleet was crewed by] more than 27,800 men included sailors, clerks, interpreters, officers and soldiers, artisans, medical men and meteorologists. On board the ships were large quantities of cargo that could be broken down into over 40 different categories, including silk goods, porcelain, gold and silver ware, copper utensils, iron implements, cotton goods, mercury, umbrellas and straw mats.
>
> - Irena Knehtl in the *Yemen Times* (September 2005), celebrating the 600th anniversary.

Interestingly, there are people in the Lamu archipelago on the Kenya coast who claim descent from Chinese traders or sailors from the fleets of Zheng He who may have been deposited there to form a trading depot, or lost from some disaster. The Sultan of Malindi gave a giraffe to the visiting Chinese fleet which survived the voyage and transportation to China in 1416, where it was used as a religious or spiritual symbol.

Emperor Zhū Zhānjī of the Ming dynasty died in 1435 and Chinese ocean trading and exploration in the Indian Ocean ceased with him. It was at this time of the height of this Chinese maritime power that the next Emperor determined on an inward policy, which more-or-less lasted until the end of Chinese imperial rule. The cessation of organised Chinese voyaging in the Indian Ocean is generally ascribed to renewed Mongolian threats from the north and west, the enormous cost of the great fleets which had not yet proved profitable, the need for greater attention to agriculture and the opposition of the conservative Confucian bureaucratic hierarchy. Zheng He was a Moslem and a eunuch and there was much concern amongst the Mandarin civil service establishment at the influence that he and his ilk were having over the Emperor. China probably had a population of more than 100,000,000 at that time and was a complex nation-state with no equal anywhere in the world until the 19th century.

The tentative venturing of the European maritime explorers and traders, whom we rightly see as heroes, would have cast a very small shadow beside the giant umbrella of the Chinese had they wished to maintain an overseas trading hegemony.

The industrial might of civilised humankind for three millennia was located in Asia and the Indian Ocean was one of the main pathways for the commerce which served this enormous engine of production. Eurocentric history and opinion seldom recognises the shear volume and sophistication of the industry and commerce of the nations surrounding these seas, perhaps because there was nothing like it in Europe, apart from the Romans, until the seventeenth century. Any view that assumes that eastern Africa was a *terra incognita* to seatraders until the rise of Islam is one which is blinkered. People from Egypt, India and China did not build permanent cities on the East African shores principally because there was insufficient incentive. The gathering of ivory to await the monsoon-driven traders once a year could easily be left to local African chiefs and their organisation.

However, although Arabian or Persian Gulf princes did not establish colonies during the first centuries of the Christian era, a Swahili people were being born. The *Periplus* states quite clearly that merchants and sailors were intermarrying with locals.

It was only after the gold trade with the interior of Zimbabwe flourished that Arabian colonies were firmly established on the East African coast within an Islamic cultural umbrella. Even so, the source of the gold was jealously guarded and there are no records available, from any source, which describe the ports of the Sofala coast or the location of Great Zimbabwe and its gold fields. It is all second and third hand information, and vague. It is not clear to me that there ever was an ancient port called Sofala.

The Portuguese named one of the small Swahili settlements on the Mozambique coast, Sofala, but it has long been my feeling that it was the name of the land as a whole, and there was no Arab or Swahili town of that name. There has never been a record of a specific town in Arabic writings. The name itself is probably derived from the Arabic word, *safala*, meaning 'low lying' and the whole coast from the Zambezi to Cape Correntes is low-lying.

Felix Chami traces the continuing contact between East Africa and the great nations of the northern hemisphere until the conversion of all Swahili towns and cities to Islam by the end of the 13th century AD. He wrote as a conclusion to his paper, *The Graeco-Romans and Paanchea/Azania* :

> The new era came with the adoption of iron technology in the early centuries AD or slightly before. The communities of East Africa grew in size and now trading with the Romans and people of Arabia. People are more cultivating and more settled. These are people trading with the Romans via the Red Sea with their capital identified by the Periplus and Ptolemy as Rhapta. The culture and the economy of these people spread quickly to the deep interior and as far south as southern Africa. The core of these communities may have been in the Rhapta-Rufiji region where many sites of that time period are found some with remains of trade goods from the Mediterranean region.

The emergence of the Swahili culture has always been something of an emotive subject. Before archaeology did more than scrape at the surface of early East African history, opinion was much influenced by the state of East Africa as described in contemporary European literature, mostly Portuguese until the 18th century and British in the 19th. The Portuguese found that the rulers of the several well-organised and independent city-states from Moçambique Island to Malindi were clearly part of formal Arab government and religious culture, calling themselves Sultans or Sheiks. The Portuguese themselves broke up the courts and governing structures of the ruling dynasties in their efforts to subjugate those Sheiks who refused clientship, and in appointing local Swahili leaders who were pliant.

Portuguese literature, mostly reports to home authority, was mainly concerned with relations with the hierarchy. Therefore, there was little description of the ordinary citizenry. Similarly, British literature described the influential people with whom they had dealings, or in the case of explorers they described the Zanzibaris or similar who were their principal servants and factotums, or the captains of slave-trading caravans and depots in the interior. In the British time these influential Arabs and Swahilis were of the period after the Omani Arab conquest of East Africa. Zanzibar and the coast was nominally the sovereign territory of the Sultan of Zanzibar and his aristocracy, and the Sultan of Zanzibar was of the same ruling family as the Sultan of Oman. Those with Omani ancestry were the majority of the ruling class who also controlled the plantations and the extensive slave trade. Neither the Portuguese nor the British told us much about the ordinary 'natives' of the coast.

Populist historiographers such as Basil Davidson and Ali Masrui, who published books and presented major TV documentary series in the 1970s and 1980s were unashamedly anti-European and glossed over the detailed structure of East African coastal society, drawing a picture of an almost wholly indigenous African Civilisation with minimal heritage from Arabia or elsewhere. The native Swahili, it seemed to them, had been attracted by Islam and taken from it what they needed whilst their indigenous African culture was hardly affected. Their historiographical distortions were risibly transparent but only to those who had a general knowledge of the larger historical picture. A ridiculous parallel would be to claim that the industrial cities of South Africa in the 20th century were the product of an indigenous African culture. Racist bias from whatever direction serves a poor master.

Post-war professionals, Chittick and Kirkman in East Africa and Somers and Garlake in Rhodesia, laboured over the evidence which was most obvious. They began a balanced understanding of the effects of the Indian Ocean system and who was involved in it. It is Chami's generation who have been able to get to grips with greater depths. Now is an exciting time!

Chami in *Kaole and the Swahili World* :

> The Swahili culture has been defined as all archaeological sites along the East African coast dating from AD 1250. Previous scholars have seen the Swahili tradition of East Africa to date back to the 8th century AD. However, recent studies have shown the Swahili culture was established along the coast of East Africa from about AD 1250 when a cultural package developed on the northern coast from about the 10th century, [and] was spread to the south.

Chami enters the discussion about the two opposing traditions regarding Swahili culture. There is the 'colonial' or 'Orientalist' view that it was an Arab culture imposed on Africans and the 'Africanist' view that it had existed on the coast for a long time before the Arabic influence appeared. The 'Africanist' school became split into the one that favoured the African base to be Cushitic (Afro-Asiatic) and that which saw Bantu-speaking Negroes as the base. If it was Cushitic, then the African component would have moved south from Ethiopia-Somalia as Neolithic pastoralists who went on further to southern Africa, founding agriculture as they travelled. If they were Bantu, they migrated from the Interlacustrine Zone to the coast in the vicinity of Kwale and spread along the coast, those who went north being the ones who first became involved with the ocean trade.

The assumption, from Swahili tradition itself, which is soundly logical, is that the language and generally urbanised society first appeared in the area of the Lamu archipelago, bordering the undoubted Cushitic-occupied lands of Somalia (Azania as the Greeks and Romans called it). These discussions proceeded in the 1990s, after my own sojourns on the East African coast and talks I had then with Richard Wilding and others.

In 2006, Felix Chami became more definite and forthright in his opinion about the peopling of the East African shores, concluding that they had been Bantu-speaking Neolithic agriculturalists from maybe 5,000 years ago. The identification of Neolithic agriculture with Cushitic language speakers is rejected because he assumed that Cushitic was a language group that evolved after contact and merging of Bantu-speakers with migrants from Arabia to the Horn of Africa probably as late as the first millennium AD. He wrote:

> It is likely that most of what has come to be identified as Cushitic/Semitic languages in Tanzania, the Horn and modern Ethiopia, is a recent influence from Arabia taking place in the early and later part of the 1st Millennium AD and hence not related to the Last Millennium BC incursions from the South ...
> Second, and related to the first, is the fact that historically and archaeologically it can be shown that people today known as "Cushitic" never settled south of Sahara ... All the records before Herodotus, 484 BC, did

suggest that the land south of Egypt belonged to Ethiopians, clearly known to be black, Negroid, people and Herodotus himself did testify to that.

Pottery seems not to have helped over much in resolving this particular issue because language and genetic evidence of race cannot be ascribed to pottery. Especially, people of a particular race can adopt more than one culture, change it and the language in common use, and change them both again. It has usually been assumed that the Early Iron Age arrival on the coast is characterised by Kwale pottery, named after a particular site in the Pemba valley of the Kwale District in Kenya, near Mombasa. Chami adds to the discussion by suggesting :

> ... the ancestral coastal tradition of what came to culminate as Swahili tradition can be traced back to the EIW [Early Iron Age] tradition, not on the Kenya coast but on the central coast of Tanzania. Archaeologists working on this area, between Kilwa and the Zanzibar channel, have found sites with continuous occupation from the last centuries BC to the 12th century AD when the Swahili culture was being formed.

In 2006, Chami was able to go back further in time through the publication of his archaeology on Zanzibar at the cave sites of Machaga and Kuumbi, with earliest dates to 3300 BC with evidence of domestication. He considered that he could write in 2006:

> In the examination of all the pottery collected from those sites discovered by the research [Zanzibar, Mafia Island and the coast], it has now been established that all of the traditions of the Rift Valley and the Nile Valley Neolithic are present on the coast and islands of East Africa.

Ancient connection between the Nile, the Interlacustrine Zone and the Tanzanian coast using the river roads are entirely logical if one examines the geography and considers the other wider problems of cultivation (with accompanying pottery), the arrival of cattle and the spread of iron technology, matters which I have discussed earlier.

However, Chami's championing of Tanzania rather than Kenya (Kwale) as the apparent centre of Early Iron Age development and the foundation of a modern Swahili civilisation smacks of possible regional rivalry. As far as I know there has been minimal archaeology of pre-Islamic sites on the Kenya coast outside of the Lamu Archipelago for many years. In my personal knowledge, a report on a potentially important site on the Kenya south coast, not far from Kwale, was provided to the Kenya Museum Service with a quantity of pottery by a professional archeological surveyor in the 1990s but lies gathering dust.

I have no doubt that people were moving with their cattle culture out of Ethiopia and Somalia along the Rift Valley and down the coast from time to time. On the coast itself their cattle would not have survived and in oral tradition and Portuguese history they are known for their raiding and pillaging of settled Islamic towns. Most modern historiographers have assumed that Cushitic people moved down from Ethiopia ahead of Bantu farmers from the Great Lakes. The evidence of Kwale pottery with firm dates in the first centuries of the Christian era seemed to fix their arrival. It is Chami's archaeology in Tanzania and especially his interpretation which now creates some controversy.

If there was Neolithic agriculture on the Tanzanian coast in the first and even second millennium BC, then there has to be an assumption that they were not Bantu. They were Nilotic and Cushitic, people who moved south from the Nile and Ethiopia rather than east from the Great Lakes. Chami creates confusion by proposing that the Bantu migration into eastern Africa occurred many millennia earlier. Genetics show that the majority of people of the northeast region were Negroid before 80,000 years ago but that has no relevance to the particular culture of coastal people in the last few thousand years.

Accepting Chami's archeology but not necessarily his speculative interpretations, it seems to me that it was indeed people with a Nilotic or Cushitic culture, or mixed, who lived on the East African coast in the first millennium BC. Chami's pottery interpretations suggest this. Bantu-speakers with Kwale pottery arrived from the Great Lakes in the first centuries of the Christian era and spread southwards, either after mixing with the existing farmers and fishermen or leapfrogging past them. There must have been an interesting confrontation of cultures. The old-established people were sedentary farmers and fishermen with a trading relationship with the Indian Ocean. The newcomers were brash and adventurous, moving outwards with energy, having been propelled by population growth around the rich Lacustrine Zone.

No doubt there were climatic factors at work, but the introduction of trans-Saharan and Nilotic trade, stimulated by the Roman Empire, must have been a potent factor too. Roman military expeditions were sent up the Nile and into the Sahara to try to control the sources of sub-Saharan trade, but they foundered in territory for which they were quite unsuited. Chami has proposed this alternative trading system as a stimulus to East African development for some years and he has begun to prove it from his pottery trails. I see the hand of trading helping to stir all major movements in the last few thousand years; trading created wealth, political sophistication and population expansion.

Until work is undertaken on the Kenya coast and along the river roads connecting that coast with the Great Rift Valley and the Nile this somewhat confusing speculation about the culture and immediately previous origins of the proto-Swahili coastal people will have to suffice.

*

At the beginning of the Christian era, Arabs from today's Yemen claimed suzerainty but that was only in respect of the seaborne trade; they had no authority on land in those early centuries. The locals were chiefs in their own land as the *Periplus* pointedly makes clear.

I have always seen the history of the East African coast and the evolution of the Swahili culture emerging with that of Islam after 800 AD, as being of an essentially feudal style, strictly limited to the trading towns or depots. There was a ruling elite who controlled the external trade and therefore the wealth of each town. This elite and feudal class valued and proclaimed their Arabian or Persian origins, but as time passed there was mixing with locals, most particularly with household slaves. Again, the *Periplus* describes this process occurring long before the Islamic period. There was no mass immigration of Shirazi Persians or Arabs, any more than there was of Portuguese in the 16th century.

Chami defines the beginning of the true Swahili cultural era with the commencement of the construction of substantial stone towns and a burgeoning of seatrading. The general adoption of Islam is illustrated by the construction of permanent and imposing mosques. This flowering began in the 13th century and it coincided, not unsurprisingly, with the rise of the Zimbabwean Empire in the south and an increasing flow of gold. It was based on new and high-value trade.

I see that the feudal tradition, emerging from a trading regime in the first millennium BC, was maintained in varying degrees of formality right up to the establishment of the conquering Omani Sultanate based on Zanzibar in the 18th century, which itself was taken over by the German and British colonial administrations in the 19th.

A pattern was set early on. Traders from the powerful and sophisticated civilisations of the northern hemisphere established connections with coastal people with whom a relationship of trust could be established. Trust is vital when the monsoon wind system means that the ships will come once a year and much of the trade goods, elephant tusks, rhino horns, tortoiseshells, aromatic gums and beeswax, must be accumulated from hunting and gathering expeditions in advance. Occasionally, the ships would not come, old trading partners passed on, and as the centuries rolled there were new people on the coast who did not know or understand the mechanisms. So there were inevitable periods when the trade failed and both sides would have been disillusioned with conflict and hiatus.

As has happened over and again all over the world, to maintain continuity permanent agents and residents of the overseas traders were established on the African coast. If this suited both sides, these enclaves grew. If there was instability, the trading communities strengthened their settlements and formally structured cities were established. Sooner or later, a local hierarchy emerged which claimed suzerainty over the enclave. The local hierarchy became integrated into the local population but naturally maintained an aristocratic posture for their ruling class. A feudal system grew. After the launch of Islam, this aristocracy aligned itself with the overseas powers who were their trading partners and their support. It was all a natural progression. Stone towns grew and when wealth accumulated, as at Kilwa, palaces and trading caravanserais were built.

When the Portuguese came in the late 15th century and established their hegemony in the 16th, they inherited an existing system and structure of trading cities. Their conflict was with the aristocratic rulers of the city states, not with the mass of African inhabitants. Much misinformation has been irresponsibly bandied about regarding the Portuguese in East Africa. Their objective, like the Arabs before and after them, was to trade and milk best commercial and fiscal advantages from the Indian Ocean trading system. Trade can only flourish in a land of peace and stability and for most of their time there was tranquillity. They were never settlers needing to conquer territory of the indigenous populace during this time.

Wherever possible the Portuguese maintained a friendly and cooperative local sheik in power and when this cooperation was not offered, they deposed the recalcitrant one. That was the way of the times. The Portuguese were ousted in their turn by the Omanis, after the brutal siege of Fort Jesus in Mombasa, and they imposed a similar regime, appointing their own sheiks in control of major towns. Some of them, such as the Masrui family of Mombasa, revolted in time and the Omani Sultan of Zanzibar established forts in the Lamu archipelago, Zanzibar and Kilwa and resuscitated the great fortress on Mombasa.

The Sultans of Zanzibar established a plantation slave economy which lasted to the end of the 19th century and its slave-raiding tentacles reached Lakes Tanganyika and Nyassa and the upper reaches of the Congo River. As colonists, the Omanis were far more successful and pervasive with their influence than the relatively insignificant Portuguese ever aspired to.

Wherever some Swahili is understood today, the Zanzibaris traded for slaves and ivory. The Omanis during the 18th and 19th centuries were exploitive colonialists wielding greater influence and harsher impact than the Portuguese. It is late 20th century racist anti-European bias which portrays the Portuguese as ruthless and plundering foreigners and the Arabs as contributing in a benign way to an indigenous African Civilisation.

* *

The literature generally available principally views seatrading on the East African coast and the emergence of a distinct Swahili culture from a Mediterranean perspective. The *Periplus* and Ptolemy's works provided a base from which to consider the coastal culture as affected by seatraders from the Red Sea. Greek and Roman geographers and historians tell their stories. The body of maritime knowledge buried in Arabic writings has been largely ignored by most historiographers, mainly because it has not been commonly available until recently. And even today it is difficult to find.

I am particularly indebted to Gerald R. Tibbetts who produced a monumental book which contains much information in a compact volume unavailable elsewhere, together with his translation of the works of ibn Majid, the famed Arab navigator and writer of the 15th century. His magnum opus, *Arab Navigation in the Indian Ocean before the coming of the Portuguese,* was published in 1971, and until then scholars could not find the knowledge he assembled and published without treading the endless steps he took. Misconceptions and misunderstandings about early Arab and other navigators and seatraders in the Indian Ocean have been imprinted in popular histories because general writers, such as myself, have only had easy access to bits and pieces, and the speculations of others before them who were similarly frustrated.

Contemporary with the *Periplus*, ± 100 AD, an Indian document incorporates descriptions of sailing in the Indian Ocean, showing both an existing literature which is not European in origin, and that Indian sailors were routinely navigating the seas. It is the *Jatakamāla of Aryā Sūra* which, apart from other technical information, describes "the courses of the celestial luminaries" to find pathways on the seas.

I have mentioned that there was something of a hiatus in general trading with the demise of the Roman Empire. It never ceased, but as the Roman engine faltered, so the economy of the civilised world slowed. Following the conquest of Arabia by Mahomet's four immediate successors in the seventh century AD, the first Caliphs, the Umayyad Dynasty in Damascus, led the Arab and Islamic conquest during the next hundred years. In that remarkably short time, an Arab hegemony was established over the whole of North Africa, the Middle East and well into the doorsteps of Christian Europe. But this empire was western-oriented and exercised by the conquest of the land. It had no serious interest in the Indian Ocean apart from the traditional trade based on the Red Sea.

Indian Ocean seatrading remained moribund until boosted during the Abbasid Dynasty based on Baghdad (758 - 1250 AD) and the Arabian Gulf ports became the important bases of shipbuilding and navigators. Trade between Egypt, and the Mediterranean, via the Red Sea, with India and the Far East, and the Horn of Africa, was now carried by Arab dhows built and largely manned in the Gulf. After 1250, Turks entered the picture, but it was Arabs who were dominant until the arrival of the Europeans with their better technology for shipbuilding and armament.

On land, the Islamic Jihad launched from Damascus and then sustained from Baghdad, swept across North Africa and it is not surprising that there was an effect on East Africa. Ethiopia was determinedly Christian and the people of that mountainous region resisted Islam for religious as well as nationalist reasons. So Ethiopia was bypassed and the Somali and northern Kenyan coasts were early objectives of direct trading and the establishment of depots. Archaeologists explored the early Arab and Persian towns and colonial depots established by a number of trading hierarchies. The Lamu Archipelago and the Kenya-Tanzanian coast have been a rich source of archaeological sites where Chittick and Kirkman pioneered and Horton, Wilding, Abungu, Kwekason, Chami and others have worked in recent years.

After 750 AD during the Abbasid Empire, Arabs from the Yemen, Hadhramaut and Oman and Persians from the Gulf began founding colonial outposts in the Lamu Archipelago and their trading depots began to appear southwards at Malindi, Mombasa, Pemba, Zanzibar, Mafia Island and Kilwa, and various places in between. Exploration went on down the Mozambique coast, past the Zambezi delta, as far as the mouth of the Save River and the two fine bays of Bazaruto and Inhambane. Sailing instructions and trading information began circulating and the safe boundary of navigation was accepted as being Cape Correntes (Cape of the Currents), almost exactly on the Tropic of Capricorn (24°S), after which the monsoon winds become fickle and the southward-flowing current could carry ships to oblivion. Ships designed to sail with the tropical winds could not face the frequent storms off Natal and the Cape of Good Hope.

Tibbetts identifies Arabic writings which described ocean voyaging and trading during the time of the Abbasid Empire, and notes that ibn Majid refers to them and used them as sources for his own work. There was the *Kitāb al-Masālik wa'l-Mamālik* of ± 850 AD which described voyaging to India and the Far East as far as Korea. The altitudes of the Pole Star were used for determining latitude. Another three Arab authors are identified as writers on ocean navigation, and they were known to ibn Majid. Although there may be some confusion about their time, it is most probably about 950 AD. They were: Muhammad ibn Shādhān, Sahl bin Abbān and Laith bin Kahlān. Ibn Majid also quotes two Persian pilots who compiled information about 850 AD: Ahmad ibn Tabrūya and Khawāsghir bin Yusuf bin Sabāb al-Ariki.

A particular characteristic of the Abbasid Empire is that it became international through economic and artistic flowering and great energy in pursuing international trade by land and sea. The empire encompassed Moslems of any nation. Persians, and other nationalities in the Middle East and the Indian sub-continent, joined Arabs as subjects of the Caliph of Baghdad, and thus it was that East African trading city-colonies and depots were founded by both Persians and Arabs and trading vessels from all around the northwestern Indian ocean were welcome. Conflict during the Caliphs' reigns was minimal and it was not until later that territorial tensions provoked sporadic warfare along the East African coast between rival cities and their rulers.

Apart from the evidence of these identified epic poems and navigational stories, Tibbetts concluded that Arabic writings on oceanic voyaging had probably begun by 400 AD, before Islam and at the end of the Roman era.

Arab navigators described the locations of various places along the eastern African shores by the latitude calculated from the altitude of the Great Bear, the Pole Star being impractical as an observation in most of the southern hemisphere. Scholars have had difficulty in identifying the places because of the inaccuracy and variations in the latitudes as described by those navigators and the difficulty in reconciling their names and names given by the Portuguese after 1500.

However, there are some which are certain, and from Kilwa southwards they include Moçambique Island (known as Masambiji or Malambuni), the Zambezi delta (usually known as variants of

Cuama or Kuwama), Sofala, Chiloane, the Save river, the Bazaruto archipelago (also known as the Boçiças islands), possibly Inhambane and more certainly Cape Correntes. Some authorities consider that places further south are identified such as Delagoa Bay and Inhaca Island and there are vague definitions of the latitude of the end of Africa where the coast swings around to the west and north.

A Turkish admiral, Kâtibi Rûmî, compiled a sailing compendium with charts in 1554, during the reign of Sultan Suleiman Khan (1519-1566). It was one of several works known as the *Mohît*, *"The Mirror of the Indian Ocean"*. Down the East African Coast there are a number of recognisable ports of call on his maps: Bandar Maqdaso (Mogadiscio), Malandi (Malindi), Mombasa, Wasini, Zanzibar, Kilûa (Kilwa), Mosambig (Moçambique Island) and others. The interpretation of his latitudes show good accuracy and it is also possible to identify the following along the southern Mozambique shore: Khôr Kuwama (Zambezi or Cuama River), Sofâla, Kiluane (Ilha de Chiloane), Sar-Nôh (Cape Correntes). I have been fascinated that place names on the eastern African coast have had integrity over at least five hundred years in the literature of diverse people.

It is noteworthy that this *Mohît* locates two places south of Cape Correntes: Waân at about 25º S (possibly the Limpopo River) and the significant Sagara at about 26º S. I accept that Sagara is Delagoa Bay, which links with archaeological suggestions that Arabs were in contact with Tsonga-speaking traders of that area before the Portuguese presence, and there was exchange of intelligence even if there was no great trade in goods.

Tibbetts has detailed the interpreted locations described by Majid on the southern Mozambique coast, and they are repeated in the Turkish admiral's *Mohît*. This clearly shows how the knowledge was common amongst navigators in the 15th-16th centuries. Tibbetts especially uses the writings of Sulaiman al-Mahri, some few years after Majid, as references to support and extend Majid's information. There is general consensus throughout that Cape Correntes was the accepted southern limit of navigation, although there are names given in different texts for places beyond. An island, Wan or Waân, is one which Tibbetts suggests is in fact the promontory at Inhambane since there are no islands between Correntes and Inhaca, but this seems unlikely since Inhambane is north of Correntes. The most important name common to all is the harbour named Bandar-al-Shajara (Sagara) which has to be Delagoa Bay

There is no doubt in my mind that Arabs explored beyond Delagoa Bay but found that the seas were inhospitable and dangerous. The winds were unreliable and could be unpredictably stormy and the land had nothing to offer. Therefore the shore of today's South Africa was not described even in the vaguest of terms and no places were named. References to the end of Africa and the changing direction of the coastline into the Atlantic may infer that Arabs had sailed there and reported on the Cape of Good Hope, but my intuition is that all pre-Portuguese knowledge of far south African geography came from the classical literature describing the Phoenician voyage at the time of Pharaoh Neccho (approximately 600 BC). Arab scholars were certainly acquainted with Ptolemy's works and ibn Majid frequently quotes 'ancient' writings and navigational lore.

Ahmad ibn Majid al-Najdī was descended from an Arabian Bedouin clan which settled in Oman and his ancestors became seafarers. His grandfather and father were both navigators and seatraders, particularly noted for their knowledge of the Red Sea, who wrote poetic treatises and Majid followed their tradition, extending his personal expertise over the Indian Ocean. He wrote voluminously and his writings somehow survived in various pockets of knowledge. This was a comparatively esoteric study to Europeans, however, and the volume of Portuguese, Dutch, English and French navigational lore on the Indian Ocean built up rapidly from 1500 onwards. But, Majid is important to a study of Arab, particularly Gulf Arab, activity in the formation of an ocean-oriented Swahili culture in eastern Africa whose tentacles spread deep into the sub-continent.

The rise of native African feudal societies and the formalisation of social structures amongst Bantu-speakers, the momentous transition to the Late Iron Age, were affected by the distant ocean traders dominated by Majid's forebears, five hundred years before the arrival of European explorers.

*   *

In 1987 I made a personal survey of the relics of the Islamic stone towns on the Kenya coast from Manda Island in the Lamu archipelago to Wasini on the borders of Tanzania, and then went on to Zanzibar. I had travelled to several of these places in a more casual way, more as an interested tourist, at different times from 1965 onwards. Descriptions of the ruined stone towns I visited are in my book, *Two Shores of the Ocean*. There are also descriptions of Sofala and Moçambique Island in 1973, before the Mozambique revolutions. In 2000, I visited Kilwa, Bagamoyo and Kaole on the same quest.

There are many reasons why Kilwa has always fascinated me. The obvious material reasons are the ruins of the fine palace on a bluff overlooking the ocean, the two mosques with corbelled roofs which still stand today, and the two later Omani Arab fortresses. But, there are important literary relics. Apart from pilotage and navigational information, provided by ibn Majid, there are several Islamic geographers or chroniclers whose descriptions of East Africa may be referred to, but Ibn Batuta personally visited the places he described. His biography, therefore, is the most authentic of all medieval descriptions of the Indian Ocean lands.

Batuta's description of Kilwa in the early 14th century has been quoted often and I add the complete text, from the Gibb translation *The Travels of Ibn Battuta*, volume II (1965):

> We stayed one night in this island [Mombasa] and then pursued our journey to Kulwa, which is a large city on the seacoast, most of whose inhabitants are Zinj, jet-black in colour. They have tattoo-marks on their faces, just as [there are] on the faces of the Limis of Janawa [native people of West Africa]. I was told by a merchant that the city of Sufala lies at a distance of half a month's journey from the city of Kulwa, and that between Sufala and Yufi [Zimbabwe], in the country of the Limis, is a month's journey; from Yufi gold dust is brought to Sufala. The city of Kulwa is one of the finest and most substantially built towns; all the buildings are of wood, and the houses are roofed with *dis* reeds. The rains there are frequent. Its people engage in *jihad* [slave trade] , because they are on a common mainland with the heathen Zinj people and contiguous to them, and they are for the most part religious and upright, and Shafi'ites in rite.
>
> Its sultan at the period of my entry into it was Abu'l-Muzaffar Hasan, who was called also by the appellation of Abu'l-Mawahib, on account of the

multitude of gifts and acts of generosity. He used to engage frequently in expeditions to the land of the Zinj, raiding them and taking booty, And he would set aside the fifth part of it to devote to the objects prescribed for it in the Book of God Most High.

The reference to the town being built of wood has sometimes been a source of confusion. Clearly Ibn Batuta was not referring to the Sultan's palace, principal mosques and other homes of the ruling class. He is describing the town of the general Swahili populace. His reference to the slaving expeditions is also noteworthy, and the way he describes the neighbouring people on the mainland as the *heathen* Zinj to discriminate them from the Zinj of the town.

Three hundred years later, the Portuguese were travelling the coast and here is a description of Kilwa Island as it was at the beginning of the 16th century from the Portuguese chronicler, Barros, and others on the first Portuguese fleets, as quoted by Neville Chittick in *Kilwa* (1974) and edited by Prof. Herbert Prins.

The city of Kilwa lies upon an island that can be circled by ships of 500 tons. The city and the island have 4000 souls, it grows quantities of fruits, it has a great deal of millet like Guinea, butter honey and wax. The hives in the trees - namely - in a jar of three 'almudes' with the mouth covered by a palm mat with holes for the bees to go in and out.

The land is very hot. The soil is red on top and there is always some green thing to be seen but there is no fresh running water. The meat is plump: there are oxen, cows, sheep, lambs, goats, and lots of fish. The sheep and lambs have no wool but are like goats. Whales circle the 'naos' [ships]. Around Kilwa Island are many small islands, all of them inhabited.

In Kilwa there are storied houses, very stoutly built of masonry and covered with plaster that has a thousand paintings. (Some of these) sturdy houses (are) vaulted in such manner that each house is a fortress. Here are found many domed mosques and one is like that of Cordoba. The greater number of the houses are built of stone and mortar, with flat roofs, and at the back there are orchards planted with fruit trees and palms to give shade and please the sight as well for their fruit. The streets are as narrow as these orchards are large, this being the custom among the Moors [Arabs], that they may be better to defend themselves. Here the streets are so narrow that one can jump from one roof to the other on the opposite side. At one part of the town the King has his palace, built in the style of a fortress, with towers and turrets and every kind of defence, with a door opening to the quay to allow entrance from the sea, and another large door on the side of the fortress that opened on the town. Facing it was a large open space where they hauled their vessels up, in front of which our ship had anchored.

In this land there are more Negro slaves than white Moors [Arabs] who work the gardens tilling the soil. The slaves wear a cloth from the waist to the knees; all the rest is naked. The white Moors, who are the owners of these slaves, wear two cotton cloths, namely one tied at the waist that reaches to the feet. Another that falls loosely from the shoulder and covers the waistband of the other. The bodies of these white Moors are well shaped, and their beards are large and frightening to see. All persons of quality carry praying beads. The people sleep off the ground on palm nets that hold one person.

Their arms are barbed arrows and well-shaped shields made strong with palm woven with cotton. Assegais like those of Guinea and better; few swords. The landing party saw four bombards, but they are not sure about gunpowder.

We saw large quantities of distilled water, and vials of good perfume for export. There were large quantities of glass of all sorts, and many kinds of cotton cloth. We saw large sacks of resin and gum, and a great amount of gold and silver and pearls.

Here they make lime thus: they pile up in a circle a lot of logs and upon them they place the corral stone and the burning logs turn the stone into lime like that made in Portugal.

There are many trees, mostly palms and others different from those of Portugal and the same on the mainland. Here they grow very sweet oranges, lemons, radishes, and tiny onions, sweet marjoram and sweet basil in the gardens which they water from the wells. They grow betel, which has a leaf like ivy and it is grown like peas each with a stick next to it; the Moors of quality eat this leaf with a kind of lime made to look like ointment. The leaves turn the mouth and teeth deep red and it is said to be very refreshing [more likely actually it was the leaf of the tamarind tree, together with the lime and the betel nut]. In Kilwa grow large quantities of peas on a kind of weed as large as the mustard plant and they pick them ripe and store them. All the gardens are surrounded by wooden fences and canes that look like a cane-break, the grass is the height of a man. There is great amount of very good cotton that is grown and sown on the island.

Here the palm trees do not bear dates, there are some that give wine [i.e. palm wine from the juice of the flower stalks] from which they also make vinegar but they do not bear coconuts which is the fruit of the others. These coconuts are as large as good sized melons with a thick skin from which they make all kinds of ropes, and inside them they have a fruit as large as a pine, which holds about a 'quartilho' of water that is very tasty to drink. Once they have taken out this water, they break the fruit and eat it. Its inside tastes like a walnut that is not quite ripe. They dry these coconuts and get from them oil in great abundance.

There are a great number of 'sambuks' as big as caravels [the Portuguese type of ship] of 50 tons; others are smaller in size. The large mtepes lie aground and are set afloat when they have to go to sea. There are no nails in these ships; the decking is lashed with palm bands and the rudder is also lashed with them. They are held fast by white resin and gum.

The Captain-major went twice through the greater part of the island and once saw at least 25 antelopes though they are hunted here. In the interior the antelopes are numerous.

Duarte Barbosa's book written in 1516 describes Kilwa, as published in translation by George McCall Theal in *Records of Southeastern Africa* (1898):

Going from Mozambique along the coast there is an island close to the mainland which is called Kilwa, in which there is a town of the Moors [Arabs]. Of very handsome houses of stone and lime, with many windows in our style, very well laid out, with many terraces; the doors are of wood very well wrought with beautiful joinery, around are many tanks of water, and orchards, and gardens with much fresh water. They have a Moorish king over them, and they trade with those of Sofala, from which place they brought much gold, which was spread hence through all Arabia Felix, as the whole country in front to Abyssinia can also be called, on account of the low lands along the sea being occupied by many towns and places of the Moors. Before the king our lord sent people to discover India, the Moors of Sofala, Cuama [the Zambezi], Angoya [Angoche] and Mozambique were all under obedience to the king of Kilwa, who was a very powerful king among them, in which town he had a great quantity of gold, because no ships went to Sofala without first touching at this island. The Moors in it are some white, some black; they are sufficiently well dressed with many rich cloths of gold and silk and cotton, and the women also with much gold and silver in chains and bracelets which they wear on their feet and arms, and many jewels in their ears. These Moors speak Arabic, and they have the religion of the Koran, and have strong faith in Mohamed. This place was forcibly taken by the Portuguese from the king, who on account of his pride was unwilling to obey the king, our lord, when many prisoners were made, and the king fled from the island, in which his Highness ordered a fortress to be built, and put them under his command and government. Afterwards he gave instructions that it should be razed, because its maintenance was neither for his service nor his gain, and Antonio de Saldanha broke it down.

Barbosa's description confirms that of Barros but also clearly describes the Arab-dominated feudal society of the elite prevailing in this wealthiest Swahili city-state at the beginning of the 16th century. Other descriptions by these early Portuguese chroniclers detail the sway of Arab-Swahili influence as far south as Cape Correntes and the hinterland of Sofala with its connections to Zimbabwe.

One advantage of the socialist regime in Tanzania following independence and the lack of commercial exploitation was the relative unchanged appearance of out-of-the-way towns. I was lucky to have the opportunity to spend time in Bagamoyo in October 1985. Bagamoyo lies opposite Zanzibar on the mainland. It was the port at the end of the up country trading routes from where slave *coffles* were shipped and became a base for European explorers and missionaries in the 19th century. I camped on the beach with my several companions and wrote in my diary:

> <u>Kunduchi to Bagamoyo.</u>   [We] *drove north along the coast in lush coastal lands: coconuts, mangoes, bananas, cashews, sugarcane in small shambas and between scattered villages. A reasonable dirt road, but hot and humid. We arrived at Bagamoyo and it was all that I had hoped for. 'Arab' houses with carved doors and wooden shutters, rough coral walls plastered with coral lime and whitewashed: all a bit mouldy and some buildings crumbling from neglect.*
>
> *I walked in a very good market with okra, beans, coconut oil, dried copra, manioc leaves and roots, onions and tomatoes and other vegetables, groundnuts, maize kernels, rice, rock-salt, dark crystals of native-produced sugar, several fresh and dried spices, tamarind seeds, sugarcane, fresh and dried fish. Lots of people in the market in Swahili dress, the women fully covered. They didn't encourage photographs.*
>
> *The 'hotel' on the seafront, though it functions as little more than a local bar and prostitutes' rendezvous, stands on the edge of a long, long beach backed by endless groves of coconut palms. Drawn up on the beach were maybe twenty big dug-out canoes, some with outriggers, and larger sailing jahazis lay out at anchor beyond the shallows. The sea here is the Zanzibar Channel, the island lies just over the horizon, and there can never be any ocean swell. There are coral reefs offshore which protect the anchorage from the adverse monsoon...*

Apart from the excavated ruins of the medieval towns and their mosques, the most interesting relics of those centuries were the *mtwepe* ships which were still being built early in the 20th century. They were extraordinary reminders of the first seatraders; images of Assyrian and Babylonian vessels of millennia ago. They were built of sawn planks but were sewn together with coconut fibre. Bow and stern were surmounted by tall figureheads, those in the bow traditionally in the shape of a camel's head. They had a single mast with a single square sail set on a yard and the sail was made of strips of plaited coconut fronds, the same material used universally for roofing. In the last centuries canvas replaced the *macuti*, but the sewn construction was used to the end. The mtwepe ships were a living connection over more than 2,000 years of seafaring.

Before the first Portuguese sailed the East African coast and began recording their visits and these lands entered European history, Arab geographers besides Batuta wrote about them. Not many descriptions have survived and some lose accuracy in poetic licence. Eric Axelson in *South-East Africa, 1488-1530* (1940), summarised the surviving abridgement of a great lost work of thirty volumes, *Murz al-Dhabab wa- Ma'din al-Jawhar* which was completed by Ali al-Mas'udi (Masudi) in 947 AD. Here is an interesting note from Masudi, explained by Axelson:

> The Zanj [negroes] were the only tribe to cross the branch of the Nile that flowed into the Ethiopian sea [Indian Ocean], Mas'udi declares, referring probably to the Zambezi. They inhabited the country as far as Sofala, which was the limit of their land. Beyond Sofala the land of the Zanj marched with that of the Waq-waqs - a people usually accepted as being the Bushmen. The neighbourhood of Sofala was wealthy, for besides being naturally fertile, it produced gold in abundance. It was not far from that port that the Zanj had their capital. It was customary among them to elect a king, whom they called Waqlimi, meaning "Son of the Supreme Lord" [or son of God]. If the Waqlimi ceased to govern justly, then he was slain, and his posterity debarred from the succession. ... The land was broken, with many mountains and sandy deserts. The region abounded in wild animals. The elephant was especially hunted for its ivory. ... In their ornamentation, Masudi declares, they wore iron in place of gold and silver. ...

Masudi sailed with Omani merchants from Sohar to Qanbalu which James Kirkman tentatively identified as the island of Pemba in *The Journal of Omani Studies* (vol. 6/1, 1983). This was their furthest port and thereafter local vessels traded southwards. Masudi stated that ivory loaded at East African ports was transshipped in the Oman and despatched onwards to India and China. Chinese trading fleets regularly reached the Indian Malabar Coast at this time.

Masudi wrote in his book, *The Meadows of Gold* (translated by Paul Lunde and Caroline Stone):

My last crossing from the island of Qanbalu to Oman was in 304 [916-17 AD]. I was on a ship belonging to Ahmad and 'Abd al-Samad, both brothers of 'Abd al-Rahim ibn Ja'far al-Sirafi of Mikān, a quarter of Sirāf. They went down with their ship in this sea, along with everyone who was with them ... At the time of my last voyage, the emir of Oman was Ahmad ibn Hilāl, son of the sister of al-Qaytāl.

It is interesting that archaeological evidence which emerged in 1996 at Thulamela near the Limpopo River confirmed oral tradition that a ruler who had failed or was incapable was ritually killed in the 17th century. This tends to show that Masudi's observations of customs in the interior of Africa was accurate.

Another geographer was abu-Abdullah Muhammad Ibn-Muhammad al-Idrisi who lived between 1099 -1166 and his compilation was from other reports and particularly from agents that he despatched to bring back information. Idrisi is notable for having been employed by the Christian King Roger II of Sicily for whom he compiled his geography and his famous map during eighteen years of scholarship. His description of the Nile region and northern Africa is particularly important.

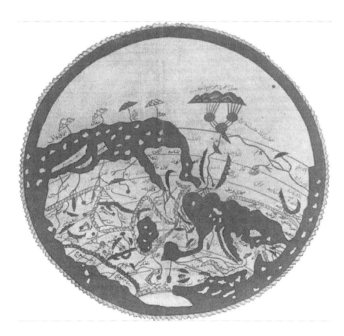

Idrisi's map of Europe and Africa as known to him in 1154 AD. South is at the top of the map and the River Nile with its source in the Mountains of the Moon is easily detected. It can be seen that he understood that the Niger River of West Africa also had the same source.

185

Abu-al-Fida (1273-1331) recorded that Arab or Swahili vessels plied the Zambezi from Sofala to Seyouna, identified with modern Sena. Dimashqui, in the late 13th century, described the people on islands in the vicinity of Madagascar, probably the Comores, who had long hair and an "easier temperament"; identified as people of Indonesian descent. Sofala is always referred to as being a source of gold, copper and iron.

I have already referred to Shaykh Ab Abdallah Ibn Batuta (1304-1368) and quoted his description of Kilwa. His narrative is filled with fascinating personal observations and descriptions of his adventures. On the East African coast he did not go further south than Kilwa, but he spent years in Arabia, Persia, India, Ceylon, West Africa, the Maldives and went to China as the envoy of the Turkish Sultan of Delhi.

Although Marco Polo (1254-1354) did not set eyes on the African shores of the Indian Ocean, his hearsay descriptions of various places and their trade was often remarkably accurate. He wrote of Zanzibar:

> Zanzibar is a large and splendid island ... They are a big-built race, and although their height is not proportionate to their girth they are so stout and so large-limbed that they have the appearance of giants. ... They are quite black and go entirely naked except that they cover their private parts. Their hair is so curly that it can scarcely be straightened out with the aid of water. They have big mouths and their noses are so flattened and their lips and eyes so big that they are horrible to look at.
> ...
> They have elephants in plenty and drive a brisk trade in their tusks. They also have lions of a different sort from those found elsewhere, besides lynxes and leopards. What need of more words? They have all their animals different from those of the rest of the world...

Marco Polo accurately described the navigational problems to the south of Zanzibar and Madagascar, warning of the southward flowing current. He wrote that Arab seatraders, having explored there, did not willingly proceed to those regions whose trade was carried out by locally-based [Swahili] sailors.

Many types of vessel which have been used in the last millennium can be seen in the western Indian Ocean today. On both sides of the ocean great dug-out canoes with outriggers are used by fishermen on the open sea. Offshore fishing from Cape Correntes in Mozambique to Cape Cormorin at the tip of India still goes on in sailing vessels whose design is as old as Ibn Majid. In the museums of Lamu and Fort Jesus at Mombasa there are models of *mtwepe* ships and, more important, fading photographs of them from the 1920s.

I have made a point, from time to time, that trade not only fosters the movement of goods and technology but it creates a movement of general and esoteric culture. The Islamic religion became entrenched as part of the Swahili culture, even though it was diluted by African native religion to a greater and lesser degree in different parts forming magic cults. Both traditional African religion and Islam were corrupted in the same way that Christianity and native religion corrupted each other to result in the grosser attributes of the West African 'juju' cults which were also carried to Brazil and the Caribbean.

But there is more to scholarship and cultural transfer than the more general evidence which is usually observed. There are many references, but my attention was drawn to a recently published book, *Sufis and Scholars of the Sea* by Anne Katrina Bang. Irena Knetle writes in her review:

> Unlike Oman, Hadramawt does not have a history of a colonial power in the Indian Ocean. Hadramawt is known for its continuous export of people to the land of the Indian Ocean, including the East African coast. They were religious scholars, traders, cultural brokers, whose impact on both recipient and home country is a topic which has aroused much interest in recent years. To them the Ocean was no barrier rather a long established arena for cultural and intellectual exchange. With them travelled goods and ideas, word of mouth, and word of writing, fashion, and habits, linguistic patterns, and seeds for new agricultural crops. They left their imprint on the place, the most notable being the religion of Islam, and absorbed cultural elements that were not Arabic in origin.

The Swahili coast does not end in Tanzania. Swahili people traded and settled in coastal trading posts as far south as Inhambane (24°S latitude) and up the Zambezi River as far as Tete. The ruler of Moçambique Island in the 15th century claimed the title of Sultan, although in practice he was probably a Sheik in the fiefdom of the Sultan of Kilwa.

There is a relatively enormous quantity of historical records on Mozambique because of the long years of Portuguese occupation. The South African Victorian historian, George McCall Theal, accumulated many references and re-told much Mozambique history in English, some of it with unfashionable bias from today's political viewpoint. Prof. Eric Axelson in the 1980s was an acknowledged expert on early Portuguese African colonial history. Dr. Malyn Newitt presented a useful summary at that time, *The Southern Swahili Coast in the first century of European Expansion*, in Azania XIII (1978). I quote a passage:

> South of the Zambezi the picture is somewhat different. It appears that when the Portuguese arrived there were a number of thriving Muslim settlements. There were at least three settlements around Sofala itself; there was an important town at Chiluane (Kilwani), and there were settlements in the Bazaruto islands and on the mainland opposite them. These are all mentioned by Majid [the Omani Arab navigator] and in addition he mentions other settlements which are more difficult to identify but which may show that Muslim traders were already active at Mambone at the mouth of the Sabi [Save]. One difference along this coast was the extent to which the Muslims had mixed and merged with the local population. The sheikh of Sofala had villages inland and had good relations with local chiefs; Muslim traders went into the hinterland to trade and the Muslims of the Bazaruto Islands appear to have lived under the protection of a chief called Moconde who cooperated closely with them. ...

Professor Paul Sinclair carried out an archaeological survey later and excavated at Chibuene in Bazaruto Bay (22°S latitude) which showed Swahili occupation from the 9th century. Chibuene would not have been in any way unique.

* *

Sanjay Subrahmanyam in *The Career and Legend of Vasco da Gama* (1997) is rightly critical of a tendency by general historiographers to describe the western Indian Ocean trading system to be an Islamic monopoly. He wrote:

> ... The very strong connotation of a term like 'monopoly' suggests a certain caution in its usage, before creating the image of a monolith where none existed.
>
> It remains true that the fourteenth and fifteenth centuries saw the expansion of a variety of forms of Islam on the shores of the Indian Ocean, and the growing presence of Muslim mercantile communities, whether in East Africa, India or Southeast Asia. But this is not the same process as that envisaged by the constructs of a 'Muslim lake' or an 'Islamic world-economy', for we should bear in mind that this Islam was as often heterodox as orthodox.

This should, of course, be obvious and the same criticism may be made of those historical writings which have branded the Portuguese and later European colonial powers with a kind of universal or monolithic style. Or indeed of the native Africans along the Indian Ocean shores. It is a common temptation to simplify, and it has to remembered at all times that the Indian Ocean is something more than a large lake. Its physical size apart, the climatic variations and weather systems divide it in time seasons and by latitude into different navigational zones needing different ships and techniques of seamanship. The geography of its littoral varies from lush rainforest to total desert. And its shores have always been inhabited by a great variety of people for the longest time of any place on this planet.

In the fifteenth century the rulers of independent trading city-states along the eastern African coast from the Red Sea to Sofala were without exception Moslems. The Arabian conquest of north Africa and the Middle East, and their imposition of Islam as the religion of the conquerors had occurred over the previous centuries. It may seem that this was an Islamic hegemony, but it is rather that there was an Arab hegemony and Arabs carried Islam with them. The spread of the words of the Prophet were indeed a motive, or a professed motive, but the objective was always trade. This hegemony was not the rule of a monolith, it is that the trading states of northern Africa, from the Indian Ocean to the North Atlantic shores of Morocco, and the cities scattered down the eastern African coast, were headed by people of Arabian origin or descent who to a greater or lesser degree practised Islam in one or other of its forms.

In the same way, the short-lived Portuguese control of the western Indian Ocean which ended in 1600 has been suggested to have been promoted by the desire to ally themselves to the Christian kingdom of 'Prester John' in Ethiopia. The spread of Christianity was a motive, piously professed, but their principal objective was always trade.

SEVEN : *A BEAUTIFUL IVORY BANGLE*
The rapid colonisation by Bantu-speaking mixed agriculturalists.

KwaZulu-Natal is a soft and lovely land lying between the high Drakensberg escarpment buttressing the southern African plateau and the warm Indian Ocean. KwaZulu-Natal has a moderate summer rainfall and a temperate, sub-tropical climate. There are many perennial rivers flowing from the Drakensberg to the sea and over very long time people have found a good life there.

    The remains of Early and Middle Stone Age people have been found, especially Middle Stone Age from the 'pulse' in the warm period, maybe 125,000 years ago. The Border Cave archaeological site in the Lebombo Mountains on the modern border between KwaZulu-Natal and Swaziland is one of the more important in Africa, and the world, with long periods of occupation at least from that time.

    Also in Swaziland, just north of KwaZulu-Natal, there is what some claim to be the earliest clearly identified mine in the world. On the peak of Ngwenya Mountain, high above the surrounding valleys, people mined haematite (red ochre) about 45,000 years ago (also coincident to the beginning of southern African rock-art). When this rich iron ore was ground to powder and mixed with animal fats it produced an attractive red-coloured cosmetic and insect barrier. Modern people, all over eastern-southern Africa, loved to smear themselves with similar mixtures, the colour depending on local earths or mineral ores which became traditional amongst them.

    Red ochre has been proven to have been used for decoration at the Blombos Cave site in the Cape from about 75,000 years ago. It was also used in Europe by Cro-Magnon people so one is easily tempted to think that the use of red ochre began long before the modern out-of-Africa migration. Manganese oxide (a black material easily powdered) was also popular.

In the Drakensberg Mountains of KwaZulu-Natal there are thousands of San-Bushman paintings of variety and beauty at hundreds of sites. For this reason the area has been declared a World Heritage Site and the volume of Drakensberg Late Stone Age paintings alone has been said to exceed, at a high level of artistic and technical merit, the total of all European examples. Remnants of San-Bushmen hunter-gatherers were still living a traditional lifestyle there in the last part of the 19th century.

About 250 AD, migrants arrived on the coast from the north. They had iron technology and their pottery was related to that of Kwale near Mombasa in Kenya. The similarity between pottery style and design from sites in seaside KwaZulu-Natal such as Mzonjani and Enkwalini, Silver Leaves in the Mpumalanga (eastern Transvaal) lowveld, coastal southern Mozambique, notably at Matola and XaiXai, and East Africa at around 200-300 AD, suggests the relatively rapid migration of the eastern stream of cultivators and fishermen, unencumbered by livestock. Richard Wilding wrote, in *Shorefolk* (1987):

> After arriving at the coast down convenient corridors across the Nyika, these people [Bantu-speaking farmers] fanned out and filtered along the coast. Along the Mozambique coast, their pottery has been found in profusion. It is dated by comparison with the material at Kwale at the earliest and Kilwa and Manda at the latest, and has been found in some profusion....

Dr. Tim Maggs in his review, *Iron Age South of the Zambezi*, in *Southern African Prehistory and Palaeoenvironments* (1984) wrote:

> The ecological pattern of Matola sites [similar pottery to Kwale] in KwaZulu-Natal is significant in terms of the present evidence for very rapid initial expansion of the EIA [Early Iron Age] down the east African coastline. Settlements scattered over some 3200 km from Kenya to southern Natal may be within 150 years of each other. Movement at anything approaching this speed would seem to require special economic circumstances, a

condition supplied by the Natal ecological model and perhaps applicable to the coastline further north as well.

South of Mozambique there is no archaeological evidence that the coastal migrations during the first centuries of the Christian era were other than those of Bantu-speaking Early Iron Age fishermen and farmers with Kwale-style pottery. There has been extensive archaeological surveying and the detailed analysis of particular sites, mostly middens on or near the seashore and in river valleys. As Gavin Whitelaw of the KwaZulu-Natal Museum in Pietermaritzburg pointed out to me as recently as 2004 there has been no evidence that cultivators other then those of the Iron Age ever occupied these southeastern coastal lands. If Neolithic Cushitic-speakers or Nilotic pastoralists penetrated south of about 20° south latitude they have left no evidence of their presence. From present evidence, it seems sure that the first Negroid colonists of southern Africa were of the Early Iron Age from their origins in the Interlacustrine Zone of Central Africa, the East African coast and West Africa via the Atlantic corridor.

*   *

Dr.Tim Maggs kindly invited me to an archaeological seminar at the Natal Museum in Pietermaritzburg in 1989. At that time Archaeological discoveries in KwaZulu-Natal were defining the outlines of the Iron-Age in southeastern Africa. A number of papers were discussed. Since then, greater detail and clarity has been obtained, but the general outline of the prehistory of the region was being established.

Dr. Aron Mazel expounded on the Late Stone Age, Leonard van Schalkwyk described his Early Iron Age sites in the Thukela Valley and Gavin Whitelaw told the story of exciting finds at the new Inanda Dam near Durban. Work in the neighbouring Transkei, south of KwaZulu-Natal where the Xhosa-speaking group of Nguni live, was described showing correlation in the early centuries of the Iron Age.

Aron Mazel provided a review of recent work on the Late Stone Age of the period from about 30,000 to 500 years ago in the Thukela Basin, the low-lying confluence of several valleys leading to the Thukela River in the heart of KwaZulu-Natal. There was another site where work was recently completed on the Mhlatuzana River near the motorway between Durban and Pietermaritzburg.

His discussion brought the broad band of time around 30,000 years ago to my attention once again. In KwaZulu-Natal there had been changes then which archaeology was beginning to clarify with increasing precision. It was when the Middle Stone Age changed to Late, with a switch to finer techniques in the manufacture of implements such as spear tips and the appearance of carefully worked ostrich egg-shell jewellery.

Mazel pointed out that there were environmental changes and evidence of social change at that time and one could draw conclusions that one influenced the other. After about 30,000 years ago in the Thukela Basin itself there seemed to have been a hiatus in occupation and people did not resettle the area until 10,000 years ago. This was related to the particularly cold period of several thousand years during the last Ice-age. The Sahara, in the geographical African mirror, was devoid of evidence of human occupation during the same period.

Populations grew steadily after 10,000 years ago, as they did in many parts of the world, and Mazel could see possible changes in social structuring with specialisation into different groups. Pottery, as in the Cape of Good Hope and Namibia, appeared about the time of Christ before Bantu-speaking migrants arrived. This suggests good proof of the widespread occupation of southern Africa by Khoi herders who were absorbed by the Iron Age Bantu-speaking Negroes. Although Aron Mazel did not speculate, presumably because he had no fossils to prove it, my immediate conclusion was that the pottery showed that there were sheep-herding Khoi in KwaZulu-Natal then.

The Khoisan seemed to have abandoned occupation of higher altitude sites after the time of Christ and moved towards the coast, coincident to the arrival of Early Iron Age farmers and fishermen. This suggested that they became clients or lived in symbiosis, exchanging hunted meat and skins for grain, ironware and maybe later preyed on domestic animals.

At about the 15th century AD, evidence of distinct San-Bushmen culture retreated to the Drakensberg. The marvellous infusion of 'click' sounds into the sonorous Nguni language has long been accepted as the result of absorption of San-Bushmen and Khoi into Nguni society, for the special characteristic of all Khoisan languages is the extraordinary variety and spread of clicks throughout their speech. I have seen San-Bushman facial characteristics in many Nguni people as the physical manifestation of this absorption of their genes, and Phillip Tobias has shown that in some Nguni communities, as much as 60% of certain genes can be identified with Khoisan ancestry. Mazel's investigations seemed to be making about 500 years ago the time when this termination of Khoisan culture occurred in KwaZulu-Natal, except in the mountains.

Leonard van Schalkwyk described the Early Iron Age sites he had worked in the Thukela Valley. The dates were from about 600-950 AD and the communities were cereal farmers with sheep and later cattle, who settled along the rivers on suitable lands. Particularly there was association with magnetite outcrops used by iron smelters and smiths. He saw that there was cooperation between communities probably because of the rigours of the climate with cyclical droughts and crop failures. The people had to learn how to prosper in this virgin territory.

There was evidence that there was denser coverage of scrub bush at the beginning of the period which was home to tsetse fly which would have inhibited the husbanding of exotic cattle. As time passed, the bush would have been cleared for cereal farming and fuel by slowly increasing populations and conditions for cattle would have improved.

Cattle-keeping increased in those three centuries and farming methods changed. If the first immigrants into that area had no cattle, or could not sustain them because of fly-born disease before the bush was thinned out, they could have been obtained later from nomadic herders in exchange for grain and iron which was an important industry in the Thukela valley with its abundance of iron ore. San-Bushmen maintained their client or symbiotic relationship and there was considerable communication and interchange in the region.

During 8-900 AD there was dramatic change. Mixed-farming agricultural occupation throughout KwaZulu-Natal expanded into the highveld leading to the Drakensberg escarpment and other archaeological exploration has shown that this coincided with a switch to social culture dominated by a cattle-cult on lands above 1000 metres elevation. Although the reduction of bush and improved husbanding methods in the lowveld may have increased prospects for cattle, and this may have aided the changeover from cereal-orientation, a cattle-dominated socio-economy required settlement in healthy uplands. Probably new strains of cereals also arrived at that time, brought by people who had developed them in similar country to the north.

Gavin Whitelaw was entrusted by Tim Maggs with the task of excavating at the Inanda Dam site near Durban. A huge dam on the Mngeni River was being constructed to add to the supply of water for the rapidly expanding city. The valley floor had shown evidence of early settlement and since it was going to be flooded, drastic methods had to be used. Areas were cleared of bush and topsoil removed rapidly as the dam wall rose. Time was short. It was a lucky place for archaeology and several important excavations followed. There were artifacts from the Stone Age, but it was the Early Iron Age that was the target, and settlements were soon revealed. The dates obtained were from about 600-800 AD, contemporary with Leonard van Schalkwyk's sites on the Thukela, 100 kilometres to the north.

Gavin Whitelaw reported masses of pottery of different styles. There was some evidence of cattle, remains of cultivated millet and local fruits, substantial quantities of fish bones and shells, ostrich egg-shell jewellery, cowrie shells (universally admired around the Indian Ocean rim suggesting the pervasive influence of seatraders) and large quantities of ivory shavings and worked ivory. Amongst the fish bones there were identifiable remains of a mussel-cracker (*sparadon durbanensis*), a large sea fish. It is a 30 kilometre walk down the Mngeni valley to the ocean from the Inanda Dam site and the fish bones and seashells indicate that these Early Iron Age people continued their coastal fish-eating tradition.

Modern Nguni people abhorred fish which is a cultural characteristic of all eastern African cattle herders. Although there are explanations for this taboo, that fish are like snakes for example, I have yet to find a convincing one. Another explanation is that it shows links with Middle Eastern Semitic food taboos carried via Cushitic cattle-oriented people from the Horn of Africa.

There were hut circles and sheep and cattle byres at the later levels. Clay walls and stone-lined pits were unearthed which immediately reminded me of the Hyrax Hill site in the Great Rift Valley and the widespread Sirikwa Holes on the East African Kenya highlands. Leonard van Schalkwyk told me that he found a *bau* [*warri*] game board carved in soapstone associated with ± 800 AD in the Thukela valley similar to the boards carved from living rock at Hyrax Hill. There were no burials in byres at Inanda, a tradition in later Nguni cattle-oriented society, which also confirms a cultural divide around 900 AD, but there was a child's skeleton in a pot and other burials.

In addition to ivory shavings and artifacts, there was a whole elephant's tusk. A glass bead was found with a possible date of about 850 AD. That was really exciting. Assuming that it was not some random aberration, there was a clear connection with Indian Ocean seatraders that far south. The significance of quantities of ivory shavings, discarded pendants and arm-bands could be enhanced by that bead. Not only was there ivory jewellery-making at Inanda, but it had been processed in the large quantities required for trade. Manufactured glass beads have always been associated with trade goods from the Middle East, India or Europe.

Gavin Whitelaw recently provided me with a copy of his paper, *The ceramics and distribution of pioneer agriculturalists in KwaZulu-Natal*, written together with Michael Moon and published in 1996. This paper describes the state of knowledge at that time. The introduction begins :

> We recorded four Matola phase Early Iron Age (EIA) sites during a cultural resource management project in the Mngeni valley, inland of Durban. The sites, together with similar sites elsewhere in the province, represent the earliest agricultural communities in KwaZulu-Natal, dating to the fifth and sixth centuries AD.

By 1985, despite the years of civil war in Mozambique, Early Iron Age coastal sites had been identified. There were sites at Namalu, Tototo and most important, Muaconi, in the north. It seemed clear to me that there was archaeological evidence confirming both Early Iron Age migrations down this coast and subsequent seaborne exploration. Chibuene is a key site proving an early seatraders' presence, maybe as early as 800 AD. In the south there are sites at XaiXai and Matola. The latter had become a definitive site with pottery identified as being of the same culture as that of Kwale near Mombasa.

*   *

The migration of Iron Age colonists over distances of up to 5,000 kilometres in a short time was a manifestation of the remarkable settlement of parts of a quarter of the African continent by Bantu-speaking people of the Early Iron Age. There were only three possible routes for this particular migration, two of them from East Africa and one from the west. I see the fastest and most direct route along the coast from East Africa. The environment was familiar all the way until the more temperate sub-tropical lands of KwaZulu-Natal were reached. Assuming the climate 2,000 years ago had stabilised and was similar to the present, there was a band of dry savannah bush, infested with tsetse fly and unfavourable to cultivation between the lush monsoon-watered littoral and inland mountains all down eastern Africa to the Tropic of Capricorn. One stream must have felt compelled to keep to the coast.

Those that moved directly south from the Cameroon, to the west of the Congo rainforest, along the Atlantic seashores, may have reached the savannahs of northern Angola, Zambia and the southern Congo before any concerted migration down the Indian Ocean coast, and it makes good sense that they did because their hesitant or spasmodic migrations must have had their origins in Cameroon many centuries earlier.

But movement in that region would have been sluggish, being compressed through a narrow coastal corridor or squeezed through the forest itself. The agriculture which had been successful in the forest would not be successful on the savannah. They lacked appropriate food crops and animals and would have had to await the arrival of better equipped people from eastern Africa with whom they could mingle or obtain the necessary food-stocks, animals and know-how.

Accepting that the movement of Early Iron Age people was easier and faster down from East Africa, there had to be a trigger. Especially, there had to be a trigger for the rapid coastal migration and the only historical clue I have is the brief references in the *Periplus of the Erythraean Sea*. If Late Stone Age farmers had been accumulating on the East African coast in pockets of settled society promoted by seatrading Arabs from the Yemen before 100 AD, no matter how sparsely, some Bantu-speaking Iron Age farmers arriving at the coast from the Kenya Highlands may have had an incentive to move on southwards along a hospitable coastline.

Maybe, the people who moved south were not exclusively Bantu-speaking. Maybe they were a mix of the original 'Rhaptarians', whom I have written about at length earlier, and Bantu-speakers with Iron Age technology. It seems likely that the arrival of Iron Age technology from the Interlacustrine Zone on the coast, together with stimulation from the seatraders, created a dynamic new population which grew in numbers and needed space. The space most easily available was the long tropical seacoast.

But the speed of the movement remains an issue. What propelled them so fast? The new lands they passed through presumably were sparsely inhabited by coastal hunter-gatherers with whom they should not have had significant conflict until numbers built up. Was it some kind of influence, however subtle, of the seatraders: either flight from their unwanted influence, or promoted by them as potential future partners on a hitherto superficially explored coast? I believe that the several strands combining here involved sophisticated people with mixed motives and time to work them through. I remark on the *speed* of the migration, but was it really so fast? Two or three hundred years is a very long time. Maybe, as usual, there was no single motive but a combination of several affecting several groups of similar people over an adequate time span.

However triggered, I believe the speed of this coastal spearhead was possible because they were unencumbered by animals having to be husbanded through tsetse-fly belts. Or maybe they lost their animals and therefore became unencumbered. They could have resorted to reliance on hunter-gathering and fishing while on the march, keeping their agricultural, iron and pottery technologies precious until they found places to settle scattered along the way. Probably many perished.

Gavin Whitelaw in his paper has pointed out that Early Iron Age sites in KwaZulu-Natal are within easy access of iron-ore deposits. The availability of iron ore seems to be one of the primary reasons for deciding on settlements and perhaps the relatively rapid movement down the long sandy Mozambican coastline was caused by a searching for sufficient of this essential commodity for their culture. Once having mastered iron smelting and become reliant on iron tools and weapons, finding easy sources of iron would be a most powerful incentive to move on. KwaZulu-Natal provided those suitable locations in river valleys, close to the coast with its familiar seafoods, with sufficiently high annual rainfall for successful cultivation. I believe this is an important factor in considering the fast migration of the Early Iron Age from East to South Africa. Perhaps it was the dominant one.

Professor Felix Chami, ever-ready to challenge perceptions, writes in *The Unity of Ancient African History 3000 BC to 500 AD* (2006):

> A note should also be made of pottery of Early Iron Working period [Chami's phrase for the Early Iron Age] from Jenne-Jeno in Mali with upturned rims identified as carination, could pass in East Africa as Kwale Early Iron Working bowls. The pottery presented by Marks and Mohammed-Ali (1991) as of Late Neolithic of Sudan could have easily been classified as an Urewe variant of Early Iron Working had it been found in the eastern African region. Similar findings west of Lake Victoria by this author raised mixed feeling as to whether these were Early Iron Working or Akira Neolithic, which was also mixed with remains of iron smelting.
>
> The opinion here is that the finding of Early Iron Working pottery elements from Chad and Sudan, and spreading as far south as South Africa, reflects another epoch in African history, just like those of Kansyore and Narosura discussed earlier. This kind of data should be used to demonstrate a period in the African past when there was widespread contact and communication, leading to the emergence and spread of similar cultural trait that occurred during the Graeco-Roman world order and re-linked the Nile Valley tradition with that of the Sahel and East Africa. ...
>
> The spread of the Early Iron Working tradition along the coast and in southern Africa needs also to be re-examined....

Chami is engaging a wider view of those parts of Africa and a synthesis with which I can have little argument. Indeed, I see Chami's general hypothesis of widely spread pre-Christian era population of related Negroid people throughout the Sahel and into East Africa as being attractive and I have discussed this earlier. What he is mostly challenging is the concept of a particular cultural, or racial, movement into southern Africa carried exclusively by

Bantu-speaking people with distant origins in the Cameroons or the west-central African forests. He proposed that Bantu people were the ancient indigenous population of southern Africa, from at least before 3000 BC, and this is not acceptable. This stand he has taken muddies the waters of his important archaeology of the Tanzanian coast and islands, and his refreshing scholarship of Africa north of the Equator.

* *

The alternative route from East to South Africa was inland through the gap between Lakes Tanganyika and Malawi (I call it the Tanganyika-Malawi gap), or immediately to the west of Lake Tanganyika. Pottery trails, pursued by Tom Huffman, confirmed that early cultural traditions proceeded through this Tanganyika-Malawi gap. East Africa between the coast and highlands along the Rift Valley is not hospitable country. In the dry season there is little surface water and in the wet there are diseases carried by mosquitoes, tsetse flies and ticks. Therefore, people and their exotic cattle had to use the chain of inland mountains and the escarpments of the Great Rift Valley as a highway, with groups dropping off here and there. In that way, a thin scattering colonised the watered and healthy highlands of southern Tanzania, Malawi, Zambia, Zimbabwe and the South African Highveld.

The inland route for migrants was longer with varied geography and contact with earlier mixed-agriculture colonists who had preceded them into the woodland savannah of Zambia. They had to get to know how to cross disease zones and river valleys on the way. They had to feel their way past the best lands which were already settled, however sparsely. There was probably mixing and merging and territorial dispute. Dominating these various interactions and exploration, the need to care for their exotic cattle caused the inland migrants to take longer to reach the southern limits of viable farming and ranching country. Evidence from the valley sites in KwaZulu-Natal suggests that the inland migrants began arriving two to three centuries after the coastal stream.

The widespread evidence of major socio-economic change in KwaZulu-Natal at about 8-900 AD suggests the continuing arrival of new people, this time more firmly committed to cattle, coincident to population growth amongst the residents causing environmental problems in those deep KwaZulu-Natal valleys. It is at about this time that a divide in socio-political organisation is perceived, with increased clan and tribal organisation and the establishment of more clearly defined structures in eastern-southern Africa.

An obvious cause of this was the filling up of the best lands for agriculture by people which demanded structures and hierarchies to settle disputes over territory, peaceably or otherwise. Livestock became a universal part of these structures wherever it was healthy for them and the importance of cattle becomes emphasised. This change is often loosely described for convenience as the divide between the Early and Late Iron Ages.

However, it is a crucial fact that trading stimulated from the outside began to have an impact, however tenuous to begin with, on southeastern Africa at this time.

About 800 AD, in East Africa there was the excitement along the coastal strip caused by the establishment of Islamic ocean-trading contacts from Mogadiscio to Mozambique. As has been discussed in the previous chapter, this was perhaps gradual and a settled and formal hierarchical Islamic Swahili society may not have been fully established until after a few centuries, but the virility of Islamic Arabian and Persian trading depots was felt from its inception. Arab chroniclers have described not only the increasing prosperity of the seatraders' colonial towns and the emergence of the Swahili mixed-race people but also the interaction with people of the immediate interior. It was not always peaceable and Ibn Batuta (1304-1368) wrote of Kilwa.

> It's inhabitants are constantly engaged in military expeditions, for their country is contiguous to the heathen Zanj [mainland Negroes].

Richard Wilding, writing in *Shorefolk* (1987) described trade between the coastal towns and the highland interior along river-roads which commenced before the Islamic expansion and continued thereafter with what must have become regular frequency. This trade waxed and waned, of course, and its emphasis changed as centuries rolled, but its character was not altered until the 18th-19th centuries with the conquest and colonisation of Zanzibar and Mombasa by the Omanis. Indian, Arab and Swahili traders remained in the towns and the people of the interior acquired and transported the ivory, rhino horns, honey, skins, aromatic gums, rock-crystal and other produce.

Wilding also referred to the persistent 'Shungwaya myths' which told that there were aggressive people of Cushitic origin who put pressure on East Africa from a northern base and caused turmoil and migrations. Wilding in *Shorefolk* (1987):

> The story is undatable. It figures in discussions of the pre-tenth century southward movements of peoples and probably represents a phenomenon lasting very much longer. ... The basic motif is that the Oromo [of Cushitic origins] or their immediate predecessors began to unsettle the farming communities before the turn of this [first] millennium. This disruption took the form of competition for land, and violence. It persisted into the sixteenth century.

The medieval Arab-Swahili stone town, Gedi visited by many tourists in Kenya with its palace of the Sultan, several mosques, town walls and warren of alleyways, is the best undisturbed example of a 14th -16th century trading city. It was sacked on at least two occasions by Cushitic Galla invaders from the north and even occupied by them for a while. Other Arab seatrading towns such as Kilifi (Mnarani) and Mtwapa, and maybe Mombasa, on the Kenya mainland coast suffered in the same way.

The impact of seatraders on the interior of East Africa, with consequent effects on the general culture and activity of all people from the Interlacustrine Zone to the limits of Bantu-speakers' colonisation of South Africa is an area of speculation which intrigues me.

Presently, there is no local archaeological or historical support in East Africa itself for the idea of a southward migration movement about 800 AD. The evidence is the changes that archaeology have shown to have occurred in South Africa at what might have been the furthest reach of a chain of reactions. Some academics might be alarmed at claims for a trail of organised cattle-oriented tribal movements southwards, either by long-range migrations or a series of shunts, and one has to be careful of circular arguments. Nevertheless, I believe it is sensible speculation.

Clues may be found in the reverse movement of highly organised tribal groups from KwaZulu-Natal to Tanzania in the 19th century which has been described in a mass of firsthand and hearsay historical records. This series of remarkable movements northwards from KwaZulu-Natal was precipitated by what has become traditionally known as the *mfecane* by the Zulu and the *difaqane* by the Sotho. It has been likened to an Nguni Diaspora.

It was sparked by ocean trading, including slaving, at Delagoa Bay on the southern border of Mozambique, and pressures from European colonists in the eastern Cape Colony during the late 18th century. Empire-building by Nguni clan chieftains in northern KwaZulu-Natal who wished to control the increasingly important trade with the Portuguese precipitated dynastic ambitions. There is also some evidence of population growth following the introduction of maize by Portuguese traders in southern Mozambique, about which there is controversy. Henry Francis Fynn, a diarist and source of much information on the Zulus of his time, makes no mention of maize in his description of cereals and grain grown by the Zulus in the 1820s: maybe maize was adopted only by some Nguni clans. Maybe population growth was the result of disrupted society and increasing inter-clan warfare, which is to be expected.

News of increasing trade with Dutch and British settlers from the Cape of Good Hope and the commencement of territorial disputes at the interface between Europeans and Nguni-Xhosa clans in the south also sent disturbing waves through KwaZulu-Natal.

Cyclical droughts which are endemic in Africa caused famine at a time of population growth and political change and general dynastic warfare broke out. Shaka was a charismatic leader who seized the chieftainship of the small Zulu clan to which he had some hereditary claim. In a fast-moving military campaign, Shaka gained control of a confederation and welded the Zulu empire together out of all the northern Nguni clans in KwaZulu-Natal.

Clan leaders and generals who did not wish to be swept into Shaka's Zulu empire, moved north with their armies and in a matter of a few decades carved out new tribal estates in Swaziland, Mozambique, Zimbabwe and Malawi. Tribal groups of other cultures, principally the Sotho and Tswana of the western South African highveld and Tsonga of the southern Mozambique lowlands, were disrupted and displaced, resulting in years of anarchy.

The British and Portuguese colonial authorities and Afrikaner pioneer settlers became embroiled in knock-on effects and their reactions fuelled the flames. One Nguni army moved directly north along the ocean and sacked every Portuguese trading post as far as the Zambezi. Fleeing Tswanas moved into Zambia and Nguni regiments ravaged Malawi and Tanzania looking for living room. They may have raided as far as Lake Victoria.

It is an immensely complicated saga which is available for study because it happened within the historical time of early European exploration of the same territory. The story of the *mfecane*, with all its ramifications which affected settled communities over hundreds of thousands of square miles in the 19th century, is a striking model of how physical and cultural pressures from outside could cause turmoil and rapid movements over decades within the continent.

Early KwaZulu-Natal European pioneers wrote about this period. Although they were not present during the beginnings of the *mfecane*, they were present in the decades afterwards and Henry Francis Fynn, in particular, was a confidant and friend of the Zulu king, Shaka. He also had many conversations with his chiefs and generals and travelled the country widely. He personally observed the empty lands from where people had fled or where clans and clan groups had been massacred or dispersed. Some three or four thousand refugees attached themselves to Fynn as clients and he managed to obtain Shaka's permission to keep them under his protection. Fynn's diaries and papers were assembled and voluminously edited quite recently. (*The Diary of Henry Francis Fynn* 1969, edited by James Stuart and D.McK.Malcolm.) Other original sources are Nathaniel Isaacs' *Travels and Adventures in Eastern Africa* and Captain W.F.W. Owen's *Narratives of Voyages to Explore Shores of Eastern Africa, Arabia and Madagascar*.

Several early travellers in Malawi and its vicinity wrote about the activities of the Ngoni offshoot of the KwaZulu-Natal Nguni's warlike migrations and their effects in Central Africa during the later half of the nineteenth century. David Livingstone's journals have references and the respected explorer, E.D. Young, describes his firsthand experiences as well as much hearsay in *Nyassa, a Journal of Adventures* (1877).

I believe that the *mfecane* and its aftermath is the dominating historical story of southern Africa in the 19th century and vital to an understanding of the Late Iron Age in general. The continual interaction of European trading activity, cultural and territorial colonialism and indigenous counteractions, feeding on each other, caused chaotic chain reactions. The story becomes merged into conflicts between British imperialism and the Afrikaner settlers' struggle for independence and are therefore often lost in increasingly detailed European colonial history. But the essential thread stands out.

The *mfecane* may seem irrelevant to events 1,000 years and more before in East Africa, and especially when considering the difference in the numbers of people involved it may seem wrong to suggest parallels. But I believe it is a model of what happens in Africa when external pressures, particularly from foreign traders, exceeded the resilience of ancient African inertia.

*

Prolonged Islamic Arab pressure on the Axum empire in Eritrea and northern Ethiopia, which had adopted Christianity in the 5th century, resulted in its collapse in the 10th century. Axum had been a powerful stabilising force and wealthy trading empire from before the time of Christ and in its decline there had to be increasing regional instability. There are the stories of the 'Shungwaya myths'. Coincidental founding of the first foreign Islamic trading outposts or depots on the East African coast seem to me to be powerful reasons for widespread effects which could have reached South Africa about 900 AD. Parallels with the tumult and empire-building during the medieval period in West Africa, following the arrival of the Islamic invaders in the Sahel, may seem even more remote, but I see them.

Those who question the relevance of the East African 'Shungwaya myths' should study the *mfecane* and the parallels between its causes and the effects of the Islamic conquest of the Axum empire, thus precipitating Galla invasions southwards. The Galla continued sporadic attacks after the Portuguese were established in Kenya in the 16th century and entered the historical record.

In the 16th century there was another event in the historical record affecting eastern Africa. It was the strange 'Simba' invasion of Tanzania and Kenya from the south. Because this invasion has been romanticised and there have been frequent references in tourist guide books and popular writings in the British colonial period to their "eating their way" in cannibalistic fashion to the north of Mombasa, they have been derided. But I cannot escape my fascination. I see the Simba heading north in a precise preview of the Nguni *mfecane*.

The massive movements of the *mfecane*-induced invasions, which faced greater opposition and population density, have been historically traced and there is no reason why the Simba phenomenon was not similar. The fact of the break-up of the feudal Zimbabwean empire in the 15th century is not disputed and the continuing rumblings of that event could have been the underlying cause. Portuguese attempts to establish a trading hegemony on the central Zambezi by means of the ill-fated Barreto-Homen military expeditions in 1569-75, are the obvious and clearly-perceived trigger.

Portuguese records have the Simba crossing the Zambezi and sacking the trading and administrative post at Tete in 1577, two years after the failed Barreto-Homen Zambezi venture and ten years before their appearance in East African records. They sacked Kilwa and in 1589 were involved in turmoil and massacres at Mombasa during battles between the Portuguese and an invading Turkish pirate, Ali Bey, who stirred up the northern Kenya coast. They were eventually defeated and dispersed by a Portuguese-led army of local tribesmen and the forces of the Sheik of Malindi.

It is additional evidence, if it is needed, that disciplined and mobile cattle-oriented warrior people moved rapidly about eastern-southern Africa. The home base of the Simbas, whom I assume were of the Shona-speaking tribal group, remained on the Zambezi north of Tete where for many years those that stayed behind were an unruly thorn in the side of traders and missionaries. Supporting the stories of their cannibalism in Kenya, there are a number of reports of this practice by them on the Zambezi. I doubt if they were cannibals in the sense that they fed on other humans. What is more likely is that they used the parts of people, often children, sacrificed to produce 'medicine' for the blessing of warriors. This practice has been widely reported throughout Africa in modern times.

*

It is reasonable that groups of aggressive East African nomads, with a 'cattle cult' and a sharp knowledge of the trading tradition that was causing disruption and change, followed the age-old route through the Tanganyika-Malawi Gap into southern Africa during the Iron Age transition.

Professor Tom Huffman writing in *The African Archaeological Review* (1989):

> Whatever the precise origin, both Sotho-Tswana and Nguni ceramic styles probably had an EIA [Early Iron Age] Urewe tradition origin in East Africa. The distances involved do not diminish the validity of the migration itself...

Socio-economic cultural evolution similar to events in KwaZulu-Natal at the watershed of the Iron Ages occurred in East Africa at about the same time. Dr. John Sutton, Director of the British Institute in Eastern Africa at Nairobi, in *A Thousand Years of East Africa* (1990) described quite abrupt changes in the culture of two main areas of ancient settlement: the Interlacustrine Zone around Lake Victoria, and the section of the Eastern Rift Valley and its highland periphery that stretches through western Kenya and north western Tanzania.

Sutton described archaeological exploration in these two zones and reaches some conclusions. Two distinct peoples inhabited this area, Bantu-speaking farmers and Nilotic-speaking pastoralists. Very roughly, about 1000 AD there was merging of economies and cultural change, but this did not necessarily mean mixing of people or new political overlords. Bantu-speaking farmers in the Interlacustrine Zone who had some cattle (rather like the Early Iron Age people in KwaZulu-Natal) acquired more cattle and culture switched in orientation (as it did in KwaZulu-Natal). Sutton wrote:

> Whereas cattle have been kept in the high grasslands close to the Rift Valley of Kenya and Northern Tanzania for quite three thousand years, in the lush pastures of the Interlacustrine Zone their history seems a lot shorter.... Quite plausibly it was not till around the middle of the Iron Age, 1000 AD or so, that specialised herding began here.

Sutton suggested that social modification occurred not because of an invasion from elsewhere but because of changes promoted by population growth within the Bantu-speaking farmers and surrounding Nilotic-speaking herders. Whereas the two cultures were separated by language and lifestyle, they were interdependent through trade. People in the high-rainfall forests within which they cultivated small patches of land, began to clear greater areas of bush to increase their agricultural output, which then enabled them to keep more livestock. The exchange of grains for cattle between farmers in the forest and herders on the savannah plains developed and knowledge and expertise followed trade. Economic improvement and diversification provoked necessary changes in social and political structures. These changes detected by archaeology in East Africa confirm the general trend typifying the division between the Early and Late Iron Ages in eastern and southern Africa.

Sutton describes the Sirikwa livestock specialists who built defensive sunken byres (kraals), the Sirikwa Holes, surrounded by semi-permanent living huts, exemplified at Hyrax Hill near Nakuru. Unfortunately, Sutton was not ready to speculate on the origin of the 'Sirikwa' people, but does conclude that they were ancestral to the modern Kalenjin Nilotic-speakers who are cattle-oriented cultivators with a society strongly organised about circumcision age-sets. (Circumcision was the powerful coming-of-age ceremony amongst many cattle people in Africa and often related to a living symbol or 'totem'. People of the same totem from different clans, or even tribes, had close affinity. The Zulus under Shaka organised their army and labour force in circumcision age-sets.) Reading John Sutton's description of these East African Nilotic-speaking Kalenjin, one could imagine he was describing the Bantu-speaking Nguni of KwaZulu-Natal and Transkei.

In his wide-ranging review of the more important East African excavations, John Sutton also described the elaborate irrigation and terracing agriculture in the Rift Valley and feudal stone- and earth-walled chiefly capitals around Lake Victoria where there was some merging between Nilotic warrior herders and Bantu farming peoples. Again, the development of social and economic structures and the evidence of technical change in eastern and southern Africa in the last 1000 years has some uncanny parallels.

Tom Huffman has noted that pottery from Kalambo Falls, at the southerly point of Lake Tanganyika and near to the medieval Ivuna saltworks on the shore of Lake Rukwa (precisely at the Tanganyika-Malawi gap) has styles that appeared in KwaZulu-Natal. The walled villages around Lake Victoria, especially the stone walls with lintelled doorways of the *ohingas* appeared contemporaneously with the stone buildings of the Zimbabwean empire and look much the same. Agricultural terracing in Tanzania and eastern Zimbabwe was practised in the medieval period and died away in the 17th century and I have already commented on it in a previous chapter.

In East Africa, political unease resulting in warlike activity at this time of changes from the Early to Late Iron Ages, suggested by the 'Shungwaya myths', may not have had any direct effect on the people of the lakes and the Great Rift Valley, but they would have been aware of it. They may seem to have been isolated from the fall of empires in Ethiopia or the setting up of trading outposts by literate strangers with a new dogmatic religion on the coast, but over two or three centuries, rumours and the gossip of passing travellers has an accumulative effect.

Slaving by the first Islamic traders or settlers on the coast must have had serious effects on people of the interior. European pioneers were always surprised at how quickly news would rapidly travel hundreds of miles through disparate people. Archaeology has shown that there were major coincidental changes from Lake Victoria to KwaZulu-Natal. I do not believe they occurred spontaneously in isolated pockets by cultural convergence and internal evolution. The effects of organised acquisition of slaves for the Arab-dominated Indian Ocean slave trade are discussed later.

The socio-economic changes in eastern-southern Africa in the last 2,000 years are different from the previous major evolutionary jumps within the Stone Ages, because they started happening with bewildering speed. Time has speeded up since the jump to civilisation and technology began driving evolution instead of climate-induced environmental change. Through the work of archaeologists in East and South Africa, this is being demonstrated. But whether evolution became primarily technology-driven rather than environment-driven I do not believe the social principles or means of communication and interchange of knowledge, expertise and ideas changed. Violent invasions and political revolutions result in cultural degradation before the wounds can be healed and progressive change begins.

If rapidly repeated infusions of new technology and dogma continue to be imposed from outside, with exponentially increasing variety and strength, then societies become confused and degraded and change is forced because there is no time to absorb each shock before the next has started.

Physical interference from outside, typified by raiding slave-traders or their clients and agents, if massive enough, has to have the most damaging effect. Even if those efforts were short-lived and instigated for a particular reason, such as the need for manpower to drain the marshes of Mesopotamia, they precipitate turmoil and the movement of refugees, and then remain in the folk-memory for generations. A hint or suggestion that similar activity is threatened immediately revives fear and reaction. The apparent isolation and secrecy that seems to surround the Limpopo kingdoms and Great Zimbabwe, sources of real great wealth, and the spread of mythology about the southern gold trade in the medieval period, may be directly attributed to the fears of invasion and conquest that remained from folk-memory of earlier Arab or Islamic Swahili slave-raiding in East Africa.

Islamic geographers and travellers writing about the Indian Ocean seatrading system, from Masudi to Batuta, describe the towns and city-states of the eastern African coasts and always refer to the gold trade of Sofala being sourced from the interior, but none of them apparently penetrated to that source or were able to describe it in detail from secondhand knowledge. I cannot believe that an energetic and curious traveller such as ibn Batuta, who ventured as far as Beijing and crossed the Sahara, would not have got to Great Zimbabwe unless he was convinced it would be a waste of time to try.

Curiously, Batuta described the Zimbabwe people as the Limis, a name he also ascribes to people of the interior of West Africa, also a source of gold at that time. Presumably, he believed that his 'Limis' were gold miners of some remote and secret part of central Africa whose product reached the outside world through two channels; one to Sofala on the Indian ocean, and the other across the Sahara by caravan to the Mediterranean.

* *

On a visit to Pietermaritzburg in 1987, I met Gavin Whitelaw. He told me more about his recent work at the Inanda Dam site in KwaZulu-Natal. He offered to show me an example of the craftsmanship of the people who lived there at about 800 AD.

We penetrated the nether regions of the museum and Gavin unlocked a cupboard in a dark corridor. He took out a dirty-looking object and handed it to me. It was a section of a bangle, the ivory perfectly preserved even though its surface was badly stained. I suppose I had been expecting some rather crudely worked piece with coarsely scraped and smoothed surface and an awkward asymmetrical shape. After all, it was from the Early Iron Age; primitive Africa of more than a millennium ago. But it was a beautiful bangle, the surface smooth and polished and the shape was perfect. The cross-section was an extraordinarily pleasing, slim pear-drop shape with superbly precise curves.

At that time I had been travelling extensively in Africa, from south to north through the Congo, and then in Kenya and Zanzibar. I had seen numerous examples of 'native crafts' from Zimbabwe, Malawi and northwards. Bangles of various kinds were always available, mostly of metal, but some in fairly crude wooden examples. I had bought a couple of metal bangles in Zambia and Kenya as souvenirs and still wear them today, more than twenty years later. That was why I had not expected much when Gavin offered to show me the ivory bangle of more than a thousand years before. The beautiful ivory bangle took my breath away, and became for me an image of those very early years of Bantu expansion into southern Africa. And that is why I have named this book after it.

Pemba River, Kenya : definitive site of Kwale pottery (ca 200AD).

Photo by the author.

Looking into the Great Rift from the western escarpment in Malawi

The Omani fort, or *Gareza*, on Kilwa island, Tanzania.

Photos by the author.

The Thukela valley in South Africa, site of many Iron Age settlements.

Author's photo

In 2007, Tom Huffman, Professor of Archaeology at the University of the Witwatersrand, published a *magnum opus* resulting from his many years of work in southern Africa. It is *Handbook to the Iron Age, The Archaeology of Pre-Colonial Societies in Southern Africa.* It is not my intention to attempt a synopsis of this major work, for it is largely outside the scope of my theme which is the pervading effect of external trading activity on eastern African societies before the arrival of Europeans.

Huffman describes in great detail the work done by himself and his associates and students on the pottery trails within southern Africa which defines the arrival of Iron Age peoples from the north down the three routes which I have been describing earlier in this chapter. I stated that it has been difficult to identify the commencement of migrations out of eastern Africa from archaeology in that region. But, East African archaeologists have explored the settlements of Late Stone Age and Iron Age agriculturalists in the Interlacustrine Zone, and on the Indian Ocean coast. I have already mentioned the 'Sirikwa Holes' style of cattle byres in village settlements and the formation of more complex clan and tribal capitals, some of which were built of stone or with defensive earth walls. There are obvious parallels with societies in southern Africa.

Apart from Kwale and Tana River pottery of the coast, types from the Interlacustrine Zone were defined. It is these pottery types which have been discovered and intensively researched in southern Africa which provide a scenario for migrations and settlement by colonists from the north, and which Huffman has detailed in his book. The precision and enormous care which has been taken in the analysis of southern African pottery made and deposited during the long period of 200 - 1800 AD enables a matrix of information on which to build a historiography of the movement and interaction of Bantu-speaking Iron Age peoples during that time. Huffman explains:

> Because pottery was an active part of culture and a representative part of the larger style [of communal art], it can be used to recognise groups of people in the archaeological record. Two conditions must be met to increase the validity of the procedure. The ceramic style must be complex otherwise it will not be uniquely representative. And secondly, the makers and users must belong to the same material-culture group. ...
> [But, material-culture groups] usually do not represent entities defined by blood, such as lineages, totems or clans. ... Material-culture groups do not necessarily represent political organisations such as tribes or chiefdoms.

And further:

> Because language is the principal vehicle for thinking about the world and transmitting these thoughts to others, there is a vital relationship between worldviews, material-culture and language. Ceramic style and the larger design field are the result of patterned behaviour and are therefore created and learned by groups of people. ... ceramic style can be used to recognise and trace the movement of people even though their size, composition, linguistic scale and other characteristic are unknown.

Interrelated with the establishment of movement and settlement through the pottery trails, study of the remains of huts and villages, and larger towns, enabled an anthropological thesis to develop which explains the evolution of political structures, economies, social customs and religion in this vast region of sub-Congo Africa. I find it particularly fascinating in that all the people who are described by their pottery and the physical structures of their lives, and who have been genetically traced, are relatively recent colonists in a time frame and geographical zone which can be quite precisely defined. When attempting a description of Khoisan or Negro peoples of eastern and western Africa and in the pulsing environments of the Congo Basin and Sahara Desert one is viewing a world which has an almost limitless horizon backward in time. In southern Africa in the context of the Bantu immigrants, we are examining two millennia only which terminate in a wealth of firsthand historical narrative, and the geography is fairly uniform: mostly savannah and Highveld grassland.

It is cattle-keeping which he generally defined as the Central Cattle Pattern which dominates Huffman's thesis.

The early coastal stream of the Kwale pottery, which is of great interest to me, did not have cattle and their political structures appear to have been the simplest: family groups, surrounded by sufficient cultivation for comfortable living with a cultural affinity with their neighbours and a tradition of mutual assistance without hierarchical structures of control or organisation. These coastal stream Iron Age people seem to have little impact or influence on the indigenous Khoisan inhabitants. Metal working families or clans were part of this farming society which also retained contact with the ocean and a fishing-seafood regime wherever it was practical.

But, over time their pottery shows that they spread away from the shore and inhabited all of the low-lying lands east of the Highveld by about 600 AD, and at that time had begun to infiltrate higher grounds following river roads. Huffman has shown that they also adopted, or re-discovered, the use of livestock and constructed their family homesteads and hamlets in the circular Central Cattle Pattern. He has proposed that these eastern stream people with Kwale pottery had brought a Central Cattle Pattern lifestyle with them, whether they had cattle or not as decided by the environment. That is easily debatable, because they could have acquired both animals and the most suitable way of husbanding them from a somewhat later Nkope pottery central stream who became dominant in the fly-free grasslands and upland savannah, but it is an acceptable hypothesis.

From about that date, and increasingly from 700-800 AD inland cattle-keeping people from the Lacustrine Zone and the Tanganyika-Malawi Gap with Nkope branch of Urewe pottery stirred this tranquil scene. It is the evolution of their society, merging with western stream people from the Atlantic side with their Kalundu tradition pottery that provides interest, debate and excitement in southern Africa.

It is noteworthy that although there was a substantial movement of people from the northwest, on the Atlantic side, into southern Africa who took the shortest route from the Bantu heartlands, it was the cattle-cult and the eastern stream Bantu languages associated with cattle-keeping which dominated. Western stream cultural attributes are visible in the archaeology of the merged people on the Highveld of South Africa, most obviously the Kalundu pottery, but the economy of cattle-keeping and political structures following from this became dominant throughout, together with eastern Bantu languages. Noting the relationship Huffman defines above between language and material-culture, it may be difficult to reconcile this apparent paradox. Indeed, the matter seems unresolved amongst academics and remains open for further archaeological exploration and discussion. Huffman is quite clear about the need for further archaeological exploration to determine more precisely the movements of people of the Bantu culture, and domestic animals, from their ancient home to the west of the Congo Basin into East Africa, around the northern side of the Congo forests (and then southwards picking up a cattle cult on the way), and those that came directly south along the Atlantic side.

My own interpretation may be simpler because it was based on less evidence, but I see it as the basis on which the greater complexity may be tested. There are the three proven ceramic traditions associated with the three perceived streams of migration, the two linguistic groups and the two economic regimes which in turn determine the structures of homesteads. There can be no doubt that cattle mostly came down the central Tanganyika-Malawi Gap highway from East Africa. There seems to be evidence that both eastern and central stream people kept to a Central Cattle Pattern homestead style and their pottery styles both originated in East Africa.

I see that cattle-keeping is an East African economic and cultural attribute with origins in the northeast long before the beginning of the movements south eighteen hundred years ago. I have discussed this in an earlier chapter. And there is some certainty that the first herders in the region were not Bantu-speaking, and kept sheep rather than cattle. Bantu-speakers in East Africa were primarily farmers in the Interlacustrine Zone with their Urewe pottery during the first millennium BC. They acquired cattle from Nilotic and Cushitic neighbours during their expansion from their lush base around the Great Lakes and pushed outwards, sweeping up both Neolithic Negro farmers and herders and the indigenous Khoisan.

Those that then proceeded rapidly down the coast lost their cattle but may not have lost other culture associated with cattle-keeping and the Central Cattle Pattern. When they moved inland to healthy lands and met those that had moved down the centre with cattle, there was little problem in adopting a well-learned economy and culture.

The people who had moved down the Atlantic coast without cattle and who were adapted from far back in time to a forest economy and lifestyle were maladjusted to the savannah south of the Congo Basin. The savannah and grasslands were alien to them. In order to survive, they became clients to the vigorous and virile newcomers from the east with their cattle-cult and the aggressive attitudes and warrior structures that go with herding. They were converted to the Central Cattle Pattern and eastern Bantu languages associated with cattle-keeping. But, some important aspects of domestic culture incorporating pottery and layouts of living areas, and the matrilineal traditions of western forest dwellers were retained. These are female concerns and clientship involving the incorporation of substantial numbers of Western Stream women into Eastern Stream groups is indicated.

In the mergings of these three streams, there were fascinating compromises and adaptations, but the cattle-cult and the Central Cattle Pattern of lifestyle dominated. It was the powerful masculine way which was the most successful in shaping the southern Africa Iron Age, but there were also strong and enduring domestic and female forest-dwelling attributes which were kept.

Huffman had formulated the general structure of the pottery trails twenty years ago, and this pioneering work has been confirmed as time passed and more archaeological work was done. More seems necessary to clarify and confirm certain misty details. It could be that some domestic small stock, pigs and goats for example, were brought down the Atlantic side if climate variation provided a corridor some time about 2,000 years ago. The finer variation of culture resulting from a wholly cultivating, wholly herding and the infinite range within the two extremes, compounded overall by climate and all the other attributes of geography and environment is the field on which archaeology must play.

It is on this base of complex interaction between the migrants from the north, merging and combining, keeping to the general culture of the Central Cattle Pattern that seatraders on the Indian Ocean began to have their impact. Trading limited to their relatively simple needs and the more dynamic transfer and merging of cultures by clientship during the Early Iron Age in southern Africa now faced greater pressures and excitements. The northern hemisphere civilisations needed the metals which are so bountifully spread about in southern Africa, south of the Congo. The impact was severe and it continues unabated today.

Photo by the author in 1971

The ruined Portuguese fortress in the Donda Estuary, Sofala, founded in 1505, destroyed by a cyclone about 1905.

Photo : Sheila Montgomery

Zulu warriors from a large clan assembled for a ceremonial event near Stanger, KwaZulu-Natal, in 1925.

EIGHT : *THE GOLDEN RHINO AND ZIMBABWE*
Indian Ocean trading influences in southern Africa.

Previously, I have stated my understanding that Sofala, the legendary source of southern African gold, was probably a place-name for the coastal region where Swahili seatraders went to exchange northern hemisphere manufactures for gold and ivory. Transhipment took place primarily at Kilwa where the Sultan controlled this enormously lucrative trade. The southern Mozambique coast is regularly buffeted by severe cyclones and tropical ocean storm systems and there is ample evidence of modern destruction and change on this coast. Whatever settlements existed over time, where dhows came for trading, must have been quite often razed by cyclones and shifting sandy coastline. Although there has not been an exhaustive search for an original port-town of Sofala, there have been inconclusive surveys principally by Paul Sinclair in the 1970s and 80s.

    The Portuguese located a small coastal town with Swahili people which was called Sofala on the Mozambique coast, 240 kilometres south of the Zambezi delta. They occupied the estuary of the Donda River which formed the port and in 1505 founded the stone fortress of São Caetano with material brought as ballast in ships from Portugal. This fortress was destroyed by a cyclone and the old town washed into a sandbank about 1905. Sixteenth century Portuguese maps show that the estuary at Sofala was sheltered by offshore tree-clad islands which had all disappeared by the 19th century.

It is not surprising that medieval Arab navigators proscribed that coast for dhow-captains who were not familiar with the conditions. Most of the trade from Kilwa southwards was carried by local Swahili dhows based at Kilwa or Moçambique Island. When Vasco da Gama visited Quelimane just north of the Zambezi mouths in 1498, he met well-dressed and sophisticated Swahilis who told him that occasionally 'white Moors', Arabs, visited as passengers or supercargo.

Archaeologist Paul Sinclair who surveyed this coast searching for the original pre-Portuguese Arab-Swahili settlement, also surmised that maybe Sofala was the name given to different sites in the past twelve hundred years as the actual trading town was shunted about by cyclonic disasters. On the shores of Bazaruto Bay, Paul Sinclair carried out excavations in 1977 and during later seasons at a site called Chibuene, five kilometres south of the town of Vilanculos. Chibuene has become pivotal to a study of Swahili penetration along the southern African coast and the commencement of seatrading as far as 22°S latitude during the period of transition from the Early to Late Iron Age.

Apart from pottery of an East African Swahili style which could have been made locally to the order of Swahili immigrants, there were significant finds of Persian glazed wares and glass beads whose style and type correspond with material excavated in the Lamu archipelago and Kilwa in East Africa. There were also pottery sherds and glass beads which correlate to Zimbabwean sites and pottery similar to that of the last years of the Early Iron Age in KwaZulu-Natal, similar to finds at the Mngeni and Thukela valley sites. During the preliminary excavations at Chibuene only one satisfactory radio-carbon date was obtained, 770 ± 50 AD. But, all the other evidence ensures that the site was a shoreside village with clear trading connections to East Africa and Zimbabwe, and possibly to KwaZulu-Natal.

Paul Sinclair in *Chibuene - an Early Trading Site in Southern Mozambique* (1982) wrote:

> Finds from Chibuene demonstrate conclusively that southern Mozambique came within the early trading networks. They bear out the suggestion that the coastal settlements south of the Save River maintained links to the north. ...

The finds from Chibuene further suggest a possible point of entry for commodities that affected the early Iron Age societies and those of the Kutama tradition of the Zimbabwe plateau and the Limpopo valley.

At Chibuene (22°02'00"S. 35°19'30"E.) there were two identified periods of occupation. The first is as described by Sinclair at the beginning of Islamic trading penetration southwards, and the second coincides with the maximum extent of the Zimbabwean empire before its collapse in the mid 15th century. During the later phase, crucibles were found, one of which contained vestiges of gold, indicating a processing depot for gold dust traded with the interior. A grave contained a skeleton buried in what was probably the Islamic tradition.

Chibuene was clearly associated with the Manyikeni archaeological site about 50 kilometres inland where a local chiefly capital was established during the Zimbabwean period. Manyikeni is an example of an outlying base of that society, controlling a seatrading outlet, but sufficiently far away to be safe from any untoward incursions by rival or unfriendly seaborne people. I have no doubt that where there was trading in precious commodities there was also piracy. Here is a brief quotation from a specialist study prepared for the Sasol Natural Gas Pipeline Project running to Bazaruto Bay:

> Manyikeni comprises an elliptical stonewall enclosure, about 50m long and 65m across, and surrounding settlement. Originally the walls were 1.50m high and 1.50m wide, made of undressed limestone. The estimated number of houses was between 100 and 140, which accommodated a population of 150-200. The spatial organisation of the structures is indicative of social stratification between the people living inside and outside the enclosure.
>
> The archaeological excavation in Manyikeni has yielded important information about the prehistory related to the role that this site has played for the development of trade activity in southern Africa during the 13th and 17th centuries AD. Manyikeni also had links with the early first millennium coastal site of Chibuene, in Mozambique and with the interior, especially Great Zimbabwe.

This important site relevant to seatrading was first described by Graeme Barker in the journal *Azania* vol. XIII in 1978.

Lyall Watson in *Lightning Bird* (1982) reminds us that there have been a number of surprising exotic objects found in southern Africa, confirming oceanic trade links that far south. As examples, he quotes a pottery figurine found in 1901 by a German archaeologist in the mid-Zambezi which was identified by Sir Flinders Petrie as Egyptian. A crystalline statue was found by a local building-sand trader in 1963 on the Umbeluzi River in Swaziland and identified at the Department of Oriental Studies of the British Museum as being an imported Bengali image of the Hindu god, Krishna. Theodore Bent in 1896 described a Roman coin found in a native mine shaft in Zimbabwe. There are many similar examples.

Chibuene was not unique. It was not a large town and nor did it occupy a site with any particular strategic importance. There are any number of similar places either with some shelter on the mainland or on the lee of islands along a couple of hundred miles of coast from the known site of Sofala at the end of the 15th century to the Bay of Inhambane at 23°45'S. That piece of coast is low-lying and has a geological structure that results in bays of varying size opening to the north, thus providing shelter from prevailing southeasterly winds, and has reasonable access, through links along the coast, to the inland highways of the Save and Limpopo rivers. An advantage that Chibuene had as an important archeological site was that it is sheltered from the worst of the weather by its location within Bazaruto Bay.

Cyclones undoubtedly caused the absence of any fixed towns or substantial seaborne immigration until the European colonial period when Europeans built in stone and other suitable material transported in their larger and stronger ships. It is interesting that the Portuguese recorded substantial Swahili settlements up the Zambezi but none of importance on the coast south of Moçambique Island, 880 kilometres northeast of Sofala and outside the usual cyclone zone.

The fact that gold was being processed and exported from Chibuene, even in small quantities, in the 15th century before the demise of Great Zimbabwe adds to the conviction that there was no single 'Sofala' as an entrepot for this most powerful of trading engines. Ibn Majid, the Omani navigator of the 15th century, stated that there was more than one port providing a source of the gold trade in the region. He mentions Chiloane which was identified in maps of that time in the Save River delta. The Save was a major river road to the interior of Zimbabwe, more direct than the Zambezi.

I see no reason why there were not a number of relatively sophisticated trading villages scattered all along that portion of the Mozambique coast from the Pungue River estuary to Cape Correntes, which were ocean-oriented and had elements of seatrading culture by 8-900 AD. After the first traders set up their bases, these villages came and went when overthrown by the occasional cyclone and might only be found by exhaustive archaeological survey. It is likely that Sofala was indeed the Arab-Swahili name for the capital of this coastal region which was not at a particular location for any great length of time. Maybe, Sofala was the name of the country and not of a town at all.

According to Tibbetts, ibn Majid in the 15th century names the King of the gold mines, Munāmunāwī, and his kingdom is called Zabnāwīi. These two names uncannily equate with Monomotapa, the name given by the Portuguese in the 16th to 19th centuries to the king of the interior empire, and to Zimbabwe.

The significance of Cape Correntes as the southern limit for practical navigation by Arab and Swahili sailors is not because at that point there is a sudden onslaught by south-flowing currents or because of the cyclone season. The adverse current begins hundreds of miles to the north and the cyclone season is easily understood. It is southwards of Cape Correntes, at the Tropic of Capricorn, where the coast becomes dramatically inhospitable with no sheltered harbours, with the particular exception of Delagoa Bay (at 26° S), and weather fronts typical of the temperate southern ocean climatic system begin to have an effect. Unpredictable gales with famously violent seas, some from the north-east, not bound by a regular cyclical season, become more frequent the further south one sails.

According to the *Periplus*, before 100 AD sailors were traversing the northern Indian Ocean and down to the African port of Rhapta (Rufiji Delta at 8°S). I do not believe they had not explored further south than Rhapta, but I do believe they had decided there was no profit in that coast at that time. When the Islamic outsurge began in the 7th century, sailors voyaged down Africa, established themselves as far as Chibuene and Inhambane and explored further. But beyond Inhambane they found nothing to tempt them to risk their ships and their lives. It is precisely at Cape Correntes, just south of Inhambane Bay, that the coastline changes to monotonous and endless beach with pounding surf backed by high bush-covered sand dunes.

Apart from a study of marine charts, I have traversed that coast on land and sea several times and I have flown up and down in light aircraft and the change to a coast generally inhospitable to oceanic sailing vessels is dramatic.

If Islamic seatraders could get all they needed in the way of ivory, rhino horn, tortoiseshell, crystalline rock, gums, honey, wild animals and their skins and slaves from the East African coast, why should they have established trading links and depots as far south as the Tropic of Capricorn? The answer, simply, is metals. East Africa, south of Ethiopia, is not rich in easily accessible metals, but the interior of southern Africa is a cornucopia. The wealth of iron and copper ores had already been established by Early Iron Age land migrants by 500 AD. Any exploring Islamic seatrading entrepreneurs in the 8th century who stopped off along the coast as far as Inhambane Bay would have met people with an abundance of iron and copper.

Enquiries about the source of metals would have been an obvious reaction. Along that sandy and tsetse-fly-ridden coast, the answer would invariably have been that metal goods and cattle were traded with people from upcountry. Trade routes with the interior along the river highways of the Zambezi, Save and Limpopo Rivers must have existed, however minimal their use was at that time.

Sea-salt and dried fish production on the coast was well-established in the historical record and it is not unreasonable to assume that salt was a principal item of trade before seatraders brought more easily-carried and profitable goods. Salt has always been a great trade-good of the Sahara. Maybe there were specialist clans who undertook one or two journeys a year during the healthy winter season. Portering heavy parcels of salt for maybe five or six hundred kilometres would require men to become trained and accustomed to the exhausting demands of the labour. In lands where flies kill exotic pack-animals, men and women have always been the carriers; men carrying trade-goods, women bearing water and firewood to the homestead.

Earlier when discussing the location of Ophir and the land of Punt, I quoted a fascinating account of the earliest recorded 'dumb-barter' gold trading in East Africa of which I am aware. An oft-repeated example of Carthaginian trading for gold on the West African coast describes how traders would place their goods on the beach and people would come with gold and exchanged it by silent barter. It is an obvious way of overcoming a language barrier when both parties trust one another and desire the trade. The same technique was used in eastern Africa and is described in the writing of Cosmas, an Egyptian monk who travelled the Indian Ocean. Known as Indicopleustes, 'The Indian Traveller', his writing is dated to about 550 AD, which is substantially before any Islamic Arab colonial activity. There is no reason to doubt this story. G.A.Wainwright, writing in the journal *Man* in 1942, identifies the trading depot as Fazogli on the Blue Nile, close to the present Sudan-Ethiopia border. Since cattle and their meat, together with iron and salt, were important trade goods, then one may assume that the miners were cultivators and fishermen. If the rainy season is as described, from July to September, this source of gold has to be in the mountainous region of western Ethiopia as proposed by Wainwright.

The Axumite empire, cited as launching these gold-trading expeditions, declined suddenly coincident to the beginning of the great Islamic expansion and presumably these contacts were terminated. Islamic Arab and Persian seatraders, seeking gold, could not penetrate Christian Ethiopia and had to find other sources, hence their exploration of the south-east African coast and eventually exploiting the gold of the Limpopo valley and then Zimbabwe.

*   *

I have mentioned what I call the Tanganyika-Malawi Gap previously when describing the filling-up of southern Africa by Bantu-speaking cattle-herding migrants from East Africa. It is one of those great natural overland gateways that geography imposes on the movement of people. With certain changes, because it lies at a junction of the slow-shifting Great Rift Valley, it will have affected mankind since the emergence of *Australopithecus* along the Indian Ocean shore of East Africa.

Buckling and heaving of the crust along the line of the Great Rift Valley raised its floor precisely at the 'Gap' and slid the line of the Rift sideways, so there is a watershed between the southern foot of Lake Tanganyika and the northern end of Lake Malawi. Two important rivers flow away north and south, both of which form pathways which have been used by travellers through country which can be inhospitable in the dry season. The Ruaha flows northwest through healthy highlands with spectacular gorges until it joins the Rufiji and proceeds to the sea opposite Mafia Island and just 75 miles north of the seatrading centre and entrepôt of Kilwa. In the Rufiji delta is the probable site for the 'lost' town of Rhapta (as described by Felix Chami) and Kilwa was the wealthiest of the Swahili trading city-states during the peak of the Zimbabwe gold trade. The Rufiji was the East African river road which directly connected the ocean and the Tanganyika-Malawi Gap.

The Luangwa river flows south from the Gap, creating a highway down to the Zambezi valley which it joins at a historic place called Zumbo, which was the Portuguese advanced trading station and mission outpost in central Africa off-and-on for 400 years. The Zambezi itself, with its tributaries, is a great highway, providing a link between west and east and between the Congo basin and the savannahs of the south. The Luangwa flows to it from the Gap, and the Shire sends Lake Malawi's overflow down to it along the Great Rift Valley which ends at Sofala. When the Portuguese began exploring, they found trading outposts with Swahilis, Indians and Arabs at Senna and Tete, on the lower Zambezi, which could be navigated as far as the Caborra Bassa gorge, 560 kilometres from the sea.

Escarpments rise on either side of Lakes Tanganyika and Malawi, providing healthy country with good rains and besides being suitable places for farmers to settle, they form part of a great inland highway from the lushness around Lake Victoria and the East African highlands right down to the grasslands of the southern African highveld. The Congo rainforest begins immediately northwest of Lake Tanganyika and although it has been shown that cattle may have been moved westwards of Lake Tanganyika, people wishing to move easily with large herds of animals southwards from East Africa, in some comfort, are unavoidably funnelled through the Gap.

Migrations with cattle herds southwards through the Tanganyika-Malawi Gap during the watershed of the Iron-Ages (800-1100 AD) undoubtedly contributed to the expansion of populations and the occupation of the whole of southern Africa to the limit of economic agriculture of that time. The Shona-speaking group of people who today occupy most of the modern state of Zimbabwe must have moved in through the gap and merged with and absorbed existing Early Iron Age people in a style which was becoming classical Bantu-speaking tradition. Tom Huffman has stated that Shona is the only modern language which can be associated with people who may have been of the Early Iron Age and arrived in southern Africa before 900 AD.

Huffman provides evidence for the infusion of a new migration southwards from the Gap by the arrival of new pottery of the Blackburn branch of the Urewe tradition in today's KwaZulu-Natal at about 1100 AD. It may be assumed that the people who made Blackburn pottery were Nguni-speaking who were described by shipwrecked Portuguese in the 16th century.

People with another pottery branch, the Moloko of the Urewe tradition, arrived later, probably from 1300 AD onwards. They may be assumed generally to have formed the Sotho-Tswana language group. These migrating people seem to have mostly bypassed already-established societies of the Shona-speaking language group with western stream Kalundu pottery inhabiting what had become intrenched settled territories along the Limpopo River valley and on the Zimbabwean highland plateau.

These settled and established societies undoubtedly owed their complex social structures and communal cohesion and strength to an external stimulus, from the Indian Ocean. At the beginning of this book, I quoted Tom Huffman on this matter of trade and socio-political development and I repeat his statement :

> It follows that trade wealth and political complexity also correlates closely. At the one end of the scale, Southern Nguni lacked both centralised chiefdoms and long-distant trade. At the other end, the only Level-6 capitals (Great Zimbabwe and Khami) and Level-5 capitals (e.g. Mapungubwe, Kasekete, Danangombe, Bulawayo and Dzata) occurred in the Zimbabwe trading zone. When the trade moved south to Maputo, Tsonga and Northern Nguni began to develop centralised polities.

* *

The Zimbabwean Empire is certainly the best-known society founded by Bantu-speaking peoples. Its spectacular stone towns, most especially the capital city of Great Zimbabwe, are famous and have been visited by throngs of tourists. My mother visited them in the 1920s and I have camped in the shadow of the walls, clambered over the hilltop ruins and wandered about the maze of interconnected stone dwelling areas on several occasions. It is a magical place.

It is the magnificence of Great Zimbabwe that has given it so great a prominence in southern African pre-history. It was trumpeted as being unique and special, an African Civilisation to rival those of Asia or Europe. Others claimed that it was astounding evidence of Arabian or even Phoenician colonisation of central Africa. An enormous quantity of racist or politically-inspired rhetoric and commentary from both directions was expounded on these themes and I don't intend to waste space on that here.

While the often bitter populist and political arguments about the provenance of Great Zimbabwe continued, professional archaeologists were at work. Before carbon-dating was invented, dates were difficult, but eventually the time-scale was established. Great Zimbabwe was at its grandest in the 14th-15th centuries and was abandoned suddenly about 1450AD. It was the cultural centre of a sophisticated feudal empire with strong trading links to the Indian Ocean system dominated by Islamic seatraders.

Many other stone towns and villages in southern Africa were properly surveyed and excavated, some much earlier than those of the identifiable Zimbabwean period, and some later. Similar feudal and imperial societies with urban complexes within the Bantu-speaking umbrella have been identified in the Interlacustrine Zone in central Africa, particularly in Uganda and around the general area of Lake Victoria. Zimbabwe was certainly African, but it was not unique. Surplus wealth from trade stimulated these societies, as has happened over and again throughout modern human history.

Fine and powerful as Great Zimbabwe is, however, it does not equal the glory of the West African empires of the same period which flourished with greater trade links across the Islamic hegemony from Arabia to the Atlantic and Mediterranean shores.

The principal difference between West African and Zimbabwean empires, apart from the volume and variety of trade and external contact, is that the former were literate and the latter was not.

* *

Mapungubwe is a cliff-begird tableland on the South African side of the Limpopo River a short distance to the east of its junction with the Shashi where Zimbabwe, Botswana and South Africa meet (22°11.5'S. 29°27'E.).

Before World War II, investigation began into the stone ruins on the top of the tableland and other remains in the surrounding valley. Subsequent periodic archaeological exploration revealed that two communities had lived there contemporaneously in a feudal society. Amongst artifacts buried in élite graves were imported glass beads and locally-made gold wire, gold beads, other artifacts and gold-plated carved wood objects including a rhinoceros: a 'golden rhino'.

Absolute evidence of trade with the Indian Ocean was established by the presence of imported glazed ceramics and glass beads. Whether seatraders themselves visited Mapungubwe cannot be proven and it could be surmised that all the trading was carried out through middlemen in a chain down the Limpopo which ended at the coast. The first contacts for commercial gold and ivory were done by word of mouth through an existing chain of copper, iron and cattle traders and it may seem inconceivable that over the two or three centuries that Mapungubwe functioned as a feudal town, growing in wealth and stature, that Swahili seatraders did not have sufficient curiosity and spirit of adventure to travel to the sources of gold and ivory. If there were no traders visiting the Limpopo, and later in Zimbabwe, it can only be because the chiefs of those societies refused them entry. I earlier mentioned the fear of slaving. It is noticeable that the kings of structured tribal societies in the 19th century would not permit European visitors to their courts until their motives were clearly established, often after long deliberation and investigation.

The river routes from the interior of southern Africa to the Mozambique coast had to be used for exactly the same reasons that they were used in East Africa. During the summer months, the lowveld flanking the ocean teemed with tsetse-fly and mosquito borne diseases affecting man and domestic beast and during the dry and healthy winter, there was no surface water away from the few major rivers. The immediate objective for seatraders wishing to make contact with a society with organised miners and metalsmiths would have been the middle reaches of the Limpopo where rich deposits of iron and copper were worked within easy distance at Phalaborwa and Messina. There were traces of gold there too, and enquiries and incentives must have led traders onward to where gold was more readily available westwards and northwards of Mapungubwe.

The discovery of the Mapungubwe ruins caused speculation when it was publicised, but it was always overshadowed by the medieval Zimbabwean culture and empire, particularly the stone ruins of spectacular Great Zimbabwe. There are a great number of stone ruins all over southern Africa and Mapungubwe was just one of these Late Iron Age sites from the last thousand years. But the particular paradox of the 'golden rhino' and other artifacts at Mapungubwe was understood for years as some eccentric outlying frontier town attached to the wealth of gold mining and craftsmanship in Zimbabwe. Mapungubwe was a surprise because Late Iron Age Bantu-speaking people apparently had no use for gold. They were cattle-oriented semi-nomadic people; gold is soft and heavy, will not alloy to make a harder material like brass or bronze and was useless to them. Without the stimulus of external trade, fine-quality gold working should not have occurred.

There is another ancient town dominated by a fortified hilltop in the Mashatu Game Reserve, which lies on the Botswana side of the Limpopo between the Shashi and Motloutse Rivers. Approaching from the north, a long narrow tableland rears up from the plain. On the flat top, there are well-constructed stone defensive walls with neat courses made from carefully masoned stone. I met a Canadian professor of archaeology from Trent University surveying in July 1983 and he described what was there. On the summit there were traces of a number of terraces for circular huts as well as the defensive walls. Below were hundreds of hut circles. Carbon dates of about 950 AD had been obtained and maybe 10,000 people had lived there at one time or another. Later settlements up to the 19th century overlaid the original town.

He pointed across the plain to another gaunt mountain with vertical rock walls. "That was inhabited, and others, but surveying will have to wait for another time." On the southern side, there was a line across the valley where the archaeologists had dug an exploration trench. At the end of the valley there was green grass and a clump of trees with fleshy leaves which signalled the position of a perennial spring. It was the source of water which made that place habitable. It was another 'Mapungubwe' and is known as Mmamagwa. The people who lived there centuries before Great Zimbabwe were numerous and had a powerful political system extending far beyond one isolated town.

Across the Motloutse River, there are other ancient stone towns from the same era, at the cusp of the Iron-age divide. Tom Huffman told me that it could be generally accepted that at about 1000 - 1100 AD there was a series of kingdoms along the Limpopo and into today's Botswana at that latitude which belonged to the same culture group but were separated into different political entities: kingdoms or dynasties. The source of their economic power was principally mining and trade with metals and their artifacts.

In June 1996, the National Parks Board of South Africa, through the medium of a documentary series on SABCTV, announced the forthcoming public opening of a remarkable Late Iron Age town they called Thulamela, situated in the northern part of the Kruger National Park near the Pafuri gate and not far from the Limpopo. It is another remarkable stone town. It has importance, further confirming the existence of a line of sophisticated towns along the Limpopo and into Botswana, based on mining and trading, as far as the Kalahari. The cultural links are clear; there is a chiefly residence on an eminence and the king was also a spiritual leader dominating a crocodile totem cult (derived from the Limpopo), evidence of a matriarchal structure (indicating a mixed western-eastern stream origin), a separate stronghold for the king's wives and all surrounded by the stone walls of family or clan communal residences. A conservative estimate of the population of Thulamela proper is 2,000 but I would guess that it was greater in the surrounding urban and dependent agricultural complex.

In 7th August 1996, the press released more detail of Thulamela, particularly the spectacular news that archaeologist Sydney Miller had commenced excavating two royal graves with gold ornaments dating from about 1550 AD. Clearly, Thulamela was part of the Limpopo cultural and trading system but had not reached its peak of development and sophistication until after the fall of Great Zimbabwe. Obvious speculation follows that Thulamela existed as an important but minor tribal centre for several centuries because of its significant geography and sprang to greater importance when an offshoot of the collapsed Zimbabwean dynasty came to occupy it after 1450.

The royal graves excavated by Miller provided much valuable material for leisured interpretation. For example, it was found that the king, who was symbolically named Ingwe, may have been stabbed by a sharp instrument from the front before being entombed. Miller has suggested that this was according to a tradition that when a leader was perceived to have spiritually failed because of natural disasters or was incompetent because of health or age, he was ritually murdered to make way for new blood. It would seem that King 'Ingwe' was the last ruler of Thulamela before it was abandoned about 1650.

The Arab chronicler, Masudi (947 AD), when describing the people of the interior of southern Africa stated that if a chief, or *Waqlimi*, failed his people, he was ritually murdered. This indicates that not only were Arabs and Swahilis in contact with the Limpopo culture at that time, but that the traditions were well-entrenched and lasted at least for 700 years. The notorious murder of King Shaka of the Zulus by his half brothers in 1824 should be re-interpreted in the light of this evidence. Shaka had by then caused misery and chaos in his kingdom by his excesses following the death of his mother, there was a drought cycle and an army had been defeated in a raid on the Gaza Kingdom of southern Mozambique.

The modern Venda-speaking people who inhabit the region south of the Limpopo are considered to be the direct inheritors of the eastern Limpopo culture. They have similarities to, and some differences from, the modern Shona-speaking peoples who generally descend from the Zimbabwean Empire north of the Limpopo. Photographs from the late 19th century of Venda towns show a remarkable coincidence of neat stone-walled communal areas which have clear resemblance to Thulamela, Mapungubwe, the later Zimbabwean ruins, and 19th century Tswana-speaking towns within reach of the western Limpopo several hundred kilometres away. Thulamela lies near the modern mining complex at Messina and gold wire and beaten gold were found there.

For me, the 'golden rhino' of Mapungubwe is a particular symbol of the cultural confrontation between ageless African peoples who never valued gold and the civilisations of the northern hemisphere who had murdered and waged wars to possess it for thousands of years. It is notable that when Great Zimbabwe was abandoned in the 15th century, large quantities of worked gold were left behind. When the people moved away after the collapse of the state they did not carry their gold away. Subsequent Shona occupiers of the ruins had no interest in the abandoned gold.

R.N. Hall and W.G. Neal, writing in 1904, described the quantities of gold found in Zimbabwean ruins at that time and the several typical manufactures: wire in several thicknesses made up in various styles of bangles, bound on ceremonial wooden objects, woven together into 'basketwork' and the finest used as thread to embellish cotton cloth; gold beads of various sizes often etched with Zimbabwean symbols and designs; beaten gold to cover wooden artifacts and sculptures (such as the 'golden rhino' of Mapungubwe); gold tacks for fixing beaten gold; ferrules for the ends of ceremonial staffs; and fine plating on copper, bronze or iron ceremonial weapons or implements. Most of that unique, finely wrought and worked gold was taken by men who scrabbled in the ruins for booty and was lost to history.

* *

Only recently has it been established that defensively constructed hilltop towns like Mmamagwa which I visited in 1983 and Mapungubwe with its symbolic 'golden rhino' preceded Great Zimbabwe and the great imperial complex created by Bantu-speaking people. Other sites such as Thulamela were occupied both before and after Great Zimbabwe. Some estimates reckon a population of more than 20,000 at Great Zimbabwe at its peak, with all the necessary organisation and protocol of a tightly controlled and complex urban capital of a grand feudal state directly influencing people over maybe 250,000 square kilometres. In earlier publications and in his latest work, Huffman describes the complicated socio-political structure of the imperial capital of Great Zimbabwe which had six tiers or classes in its heirarchy.

Great Zimbabwe was usually seen as the capital of an imperial nation that had been developed by people coming from the north who settled in that hospitable land of healthy high plateau and within reach of rich mineral deposits. Sofala was the known entrepôt with a natural route, via the Save River and its tributary the

Lundi, to Great Zimbabwe and its associated towns on the Zimbabwean plateau. In the historical record, the Zambezi was a pathway for Swahili and then Portuguese traders to northern Zimbabwe and Zambia. However, Mapungubwe was on the Limpopo with no apparent easy access to Sofala. The idea of medieval seatrading stations as far south as Chibuene and Inhambane en route to the Limpopo was not seriously considered until the 1980s.

My assumptions that Sofala was not one fixed town, and that it was the name given to the entrepôt to the gold and ivory trade of interior southern Africa as it moved coincident to the focus shifting from the Limpopo to Zimbabwe has now been accepted. Huffman proposed in his publication, *Mapungubwe* (2005), that there were two principal locations: 'Sofala 1' in the Bazaruto archipelago and 'Sofala 2' where the Portuguese found it, moribund, at the end of the 15th century.

It is easy to forget the difficulties faced by archaeologists and historians before carbon-dating became an essential technique. It was still possible, just forty years ago, for there to be complicated arguments about the dates of the stone-walled trading towns in central southern Africa.

Roger Summers, an archaeologist in Zimbabwe in the 1960s, wrote in *Ancient Ruins and Vanished Civilisations of Southern Africa* (1971):

> It is inevitable that this division of the Iron Age [Early to Late, ± 1,000 AD] should go by the name *Zimbabwe*, since the site appears to have been the centre of cultural influences which spread throughout Rhodesia, westwards to Botswana and far away southwards.

Zimbabwe was seen, not only as the source of the external trading system, but also as the stimulus for the change to the Late Iron Age. Basil Davidson, in *Africa, History of a Continent*, could state as late as 1972:

> Further south again, beyond the Limpopo, the same kind of progress from early to mature Iron Age systems occurred with the so-called Mapungubwe Culture during the thirteenth or fourteenth century. Its peoples took over the settlements of earlier Iron Age populations -

established here in the Transvaal between AD 700 and 1000 - and built a new state (an outlier of the Zimbabwe Culture) of which little is known in detail.

But better dates and greater understanding were emerging. Peter S. Garlake, an archaeological excavator in Zimbabwe at the time, wrote in *Great Zimbabwe* (1973):

> ... it is even possible that further investigations will show that the distinctions between the first Leopard's Kopje [Limpopo-oriented society] and Zimbabwe people cannot be upheld and that they are culturally identical.

In 1977, Prof. Tom Huffman was suggesting that Zimbabwe may have been founded by people from the southeast who brought knowledge of ocean trade with them to found Mapungubwe and the other Limpopo-based towns, before moving north into Zimbabwe where gold was more prolific and accessible. The dates were now more-or-less certain and it was becoming accepted that Mapungubwe was a forerunner of Zimbabwe. Ocean trading related to gold and ivory began on the Limpopo long before the foundation of Great Zimbabwe. Huffman's hypothesis of a specific migration carrying the concept of trade was difficult to prove, but what was sure was that, as early as the 9th-10th century at Mapungubwe on the Limpopo, a structured feudal society emerged coincident to the acquisition of wealth and its accumulation through trade.

Maybe, one can simply see what happened as a movement of ideas and information, perhaps carried by a few outstanding entrepreneurs or a dominating élite clan. An infusion of people towards the novelty of the developing Limpopo River structures followed, and Huffman has pointed out elsewhere that this is illustrated in the archaeological record. D.N.Beach, a Zimbabwean academic, wrote in *The Shona and Zimbabwe 900-1850* (1980):

> The appearance of the Later Iron Age on the [Zimbabwe] Plateau south of the Zambezi can now be seen to be part of just such a local movement on a much grander scale, embracing most of southern Africa and originating on the plateau south of the Limpopo [in South Africa]. .... the Kutama [a designation coming from the Shona word 'to migrate'] peoples originated in the high country on either

side of the Drakensberg, well to the south of the Limpopo.
　... In this favourable environment, the numbers of people, and their herds of cattle, increased to the point where it became desirable to move back towards and beyond the Limpopo and the southern Zambezian plateau.

Add external trading stimulus and the creation of surplus wealth within an expanding population, with the acquisition of traded artifacts from Arabia and India, and the emergence of feudal empire is obvious. This feudal society reached its peak of organisation at Great Zimbabwe in the 14th century, but it began along the Limpopo in the 9th-10th. The concept of feudalism with an aristocratic class governing a mass of less privileged peasant farmers, cattle herders, miners, warriors, servants, slaves and clients is what distinguishes these trading states of the Limpopo and Zimbabwe from other local Bantu-speaking peoples. Possibly, the aristocracy emerged from immigrant 'Kutama' families banding together to manage the increasing complexity of the state as trade promoted surplus wealth, and from the necessity of organising their communities for armies, agriculture, mining, manufacture and constructing royal buildings.

　Clan chiefs had to evolve a separate loyalty to each other and to their acknowledged paramount chief for group security and coordinated administration. This horizontal loyalty amongst clan chiefs and their extended families, thus forming an aristocratic class, subverted the independence of clans and the simple traditional vertical interlocking loyalty within clan structures typical of other Bantu-speaking people.

　This revolution may have taken place quite suddenly in response to the needs of a situation getting out of hand. It only needs one far-sighted charismatic chief to impose the concept on his colleagues and relations which then, when it is seen to work for the benefit of the élite corps, becomes the accepted mode of government in subsequent generations. When Great Zimbabwe went into decline and the empire broke up in the 15th century, some Shona-speaking people of Zimbabwe rejected the feudal system that had ruled for centuries and reverted to a simple structure of independent clans.

　However, powerful dynasties also re-formed within the wreckage of the Zimbabwean Empire and equally complex entities with the same cultural and political feudal states emerged and persevered until occasional Portuguese and then more massive Nguni interference finally erased the vestiges in the 19th century.

Indeed, Huffman has proposed that the abrupt decline and failure of Great Zimbabwe was caused by the emergence of the rival empire based on Khami towards the west, near the modern European colonial city of Bulawayo, itself founded near the royal 'kraal' of Mzilikatse, the Nguni invader who toppled the last of the Rozwi kings or Mambos and massacred his court in the mid 19th century. Once the Ruler at Great Zimbabwe lost control of the subsidiary kingdoms scattered throughout today's Zimbabwe and the Limpopo, his capital city collapsed.

Other than Huffman, no academic that I know of has openly supposed that Great Zimbabwe collapsed so abruptly because of military conquest by an ambitious King of Khami, but I see that as a fascinating probability. If Great Zimbabwe was overrun by an invading army, its ruling aristocratic classes slaughtered and the city ravaged and pillaged it could explain the wealth of abandoned mines, metal workings and sacred objects described and catalogued by Theodore Bent in his book, *The Ruined Cities of Mashonaland* (1896) and in great detail by Hall and Neal in *The Ancient Ruins of Rhodesia* (1904). The decline of Great Zimbabwe was not a gradual or orderly procedure, it shows all the evidence of sudden catastrophe.

The physical layout of Great Zimbabwe is typical of all the stone towns of those kingdoms and minor chiefdoms which grew from the wealth generated by trade. Mapungubwe is typical of the earlier examples of this evolution. A site with a prominent hill was chosen where the king's royal court buildings were built, usually with a wall around it with easily defensible paths to the summit. The king's court was also the centre of tribal religion because the king ruled by divine right, being the living symbol of the tribal ancestors. Below the hill of the king's court there was the administrative centre, public court of justice and storage for the national grain reserves. The residences of the royal wives were separate from the king's court and the queens held their own influential court with a complementary spiritual centre. At Great Zimbabwe the queens' court is the most magnificent of the public buildings: the Great Enclosure with its massive, immaculately-constructed elliptical stone wall, decorative courses, interior circular partitions, curtain walls and dramatic conical towers.

The tightly packed traditional round mud huts of the mass of inhabitants, ordered by rank, family and status, spread away from the royal and official buildings and these were separated and defined by a maze of lesser curving stone walls. Further out were farmlands and cattle posts.

The Portuguese, about fifty years later, wrested control of the Indian Ocean trading system from the Arabs and Swahilis and attempted to re-establish the precious metals trade in southern Africa. The important explorer, Antonio Fernandes, was despatched from the Portuguese fort at Sofala to examine the interior and establish contact with all the kings or chiefs of the area who could be important sources of trade, especially gold. Fernandes carried out at least two recorded expeditions, one in 1511 and the second lasting four months in 1513. He found no trace of the Great Zimbabwe empire, meeting instead a number of local chiefs over a wide swath of today's eastern Zimbabwe, only a few of whom had sources of gold but all were aware of trading.

Outside of the newly dominant Torwa or Rozwi Empire based on Khami in the west with which the Portuguese never had commerce or intercourse, there appeared to them to have been a dispersion of power and they tried in vain to discover any kind of central authority with whom they could resume the gold trade which had flourished in the 14th century. Their favourite was the 'Monomotapa' dynasty based in the northwest on the edge of the Zambezi valley, but it had no significant sources of gold.

João dos Santos wrote in *Ethiopia Oriental* (1609):

> The Kingdom of Monomotapa, is situated in MoCaranga, which in times past was wholly of the Monomotapan Empire [Great Zimbabwe], but now is divided into four kingdoms, to wit, this of Monomotapa, that of Quitere, the third of Sedanda and the fourth of Chicanga. This division was made by a Monomotapa Emperor, who not willing or not able to govern so remote Countries, sent his son Quitere to that part which runs along the River of Sofala, and Sedanda, another son, to that which Sabia [the Save River] washeth, a river which visits the sea before the Boçiças [the Bazaruto Islands]: and Chicanga a third son to the lands of Manica ...

The historically documented Monomotapa dynasty was seemingly founded in the north of Zimbabwe by one of the sons of the last emperor of Great Zimbabwe which persisted into the late 18th century, declining over time. Territorial and trade wars with other dynastic chiefs attempting to find stability after the breakdown of empire, and disruption by the Portuguese attempting to find a stable trading partner, supporting this and then that member of the royal family, finally reduced the authority of the Monomotapa state.

Other local dynasties rose and fell and the stone-building tradition continued. For example, Nalatale is an exquisite stone village with fine decorations lying halfway between Great Zimbabwe and Khami which was abandoned before completion. There are hundreds of these relics of the dispersed Limpopo-Zimbabwean culture. Thulamela may be considered the best example of this period. Machemma, north of the Soutpansberg and west of Messina, is another good illustration of later small ruins with distinctive 'Zimbabwean' decorations lying south of the Limpopo in the Limpopo Province of South Africa. I spent a day climbing about in thick acacia thornbush exploring the site of Machemma in 1977 and described in my book, *Mud, Sand & Seas*.

Mzilikazi, and his successor Lobengula, maintained a dominant position in what is today's Zimbabwe in the later 19th century. Their Nguni-speaking Ndebele people who migrated north after the Nguni *mfecane*, terrorised the Shona-speaking group who had developed the medieval Zimbabwe trading empires.

Harry 'Skipper' Hoste, an ex-Commodore of the Union Line and a troop-commander in Cecil Rhodes' Pioneer Column which annexed Mashonaland and established Fort Salisbury in 1890, described the numerous abandoned villages and hamlets of the Shona resulting from the annual raids by Lobengula's *impis* who stole their cattle and maize crops, enslaved the young men and girls and killed whatever old people and children who could not hide. His account in *Gold Fever*, published by his grandson in 1977, is firsthand and definitive. Because of this, much oral history and folklore which could have shed light on Great Zimbabwe and the later Khami-based Torwa, or Rozwi, kingdom was lost. Hoste described the abandoned stone buildings and elaborate gold mines with deep shafts before Europeans began mining operations and interfered with the ruins, seeking gold. Between them, Ndebele and then European invading colonialists wiped away the pre-history of central Africa and its centuries of contact with the Indian Ocean trading system.

It is a tendency since the great jump to civilisation, almost an unbreakable law, that when surplus wealth accumulates in a society with common culture, clan or tribal confederation *always* follows. If the surplus continues to accumulate, empires are forged across cultures usually by conquest and several tiers or classes of society grow in a pyramid beneath the King or Emperor. When there is economic regression, empires and then tribal confederations *always* break down to the natural ethnic groups of tribe and clan, often with prolonged violence.

Despite my generalisation on the dispersions and warfare and the loss of historiographical narrative, the social structures and the trading initiative of the medieval feudal states lived on amongst some of the heirs of the Zimbabwean Empire, however repressed they had been. Tom Huffman wrote in *Snakes and Birds: Expressive space at Great Zimbabwe* (1981):

> I have been able to reconstruct a cognitive model [of Great Zimbabwe] because of the well preserved archaeological evidence and the direct continuity between Shona speakers at Great Zimbabwe and the Shona speakers of today. This historical continuity is conclusively demonstrated by documentary evidence and indicates that Shona oral history, myth and traditional values are just as relevant as the early ethnographic records.

Many years later, in his masterwork *Handbook to the Iron Age* (2007), Huffman described his reconstruction of Late Iron Age societies on the Limpopo and in Zimbabwe:

> A link between politics and settlements begins with a system of courts. Throughout Southern Africa, every settlement from the lowest homestead to the highest capital had a court where men met to discuss political matters and resolve disputes. ...
>
> The smallest chiefdom on record only had three court levels (homestead, ward and chief), but some historically-known Shona and Venda groups had four levels (homestead, ward, petty chief and senior chief), while a few such as the Ngwato [in Botswana] Zimbabwe

> Ndebele and Zulu were characterised by five court levels. ... In the past, Great Zimbabwe and Khami were at the top of six-level hierarchies.

*   *

In the 19th century, Europeans found that Tswana-Sotho society on the highveld interior of South Africa continued to favour large hilltop towns and they were avid traders. When Moshoeshoe gathered together fractured southern Sotho clans in Lesotho after the ravages of the Nguni *mfecane*, he sought out flat-topped mountains as his strongholds.

Contrarily, the Nguni of KwaZulu-Natal, untouched by Great Zimbabwe, never built fortified hilltop towns or citadels. The great Zulu kings, Shaka, Dingane, Mpande and Cetshwayo, shunned mountaintops and created vast circular or elliptical towns for thousands of people. They were laid out geometrically in a formal style of immaculate beehive-shaped grass-thatched huts within a perimeter wooden stockade or *kraal*. Zulus were conquerors who feared nobody and did not need defensive fortresses on the top of hills with all the attendant inconvenience.

Tom Huffman has described, in *Southern Bantu Settlement Patterns* (1986), how hilltop towns evolved during the pre-Zimbabwean Limpopo trading period, declined with their exhaustion and were revived again because of military pressures after the Nguni *mfecane*.

Comparisons between 12th century Mapungubwe and 19th century Kaditshwene illustrate this. John Campbell, a missionary writing in 1820, described Kaditshwene. It was then a fine stone town of more than 10,000 people on a flat-topped tableland in the western Transvaal and one of several which still flourished. He wrote:

> Every house was surrounded, at a convenient distance, by a good circular wall. Some were plastered on the outside and painted yellow. One we observed painted red and yellow with some taste. The yard within the inclosure belonging to each house was laid with clay, made as level as a floor, and swept clean, which made it look neat and comfortable.

In *Twenty-Five Years in a Waggon*, Andrew Anderson wrote of the Tswana-speakers inhabiting the Transvaal-Botswana border, near the western Limpopo, in 1887:

> They are very expert in metal, melting the ore for the manufacture of ornaments, assegais, Kaffir picks and such things as they require. They also make very neat mantles, karosses and other kinds of materials for the women, the men being the tailors and dressmakers for the tribe. Time being no object, their work is beautifully executed. ... They are also very fond of music; they make various kinds of instruments which produce pleasing sounds. ...
>
> The interior of their huts and yards outside where they cook, which are surrounded by a high fence made of sticks, are kept remarkably clean and tidy, and their iron utensils also receive the best of attention. Many of these Bechuana are rich in cattle, sheep and goats. They have their cattle-posts away in the bush, where the stock is looked after, cows milked, and once or twice a week a pack-ox is loaded up with skins of milk and taken to the kraal for use. These "viehposts" are in charge of their slaves, called Vaalpans. They are the Bushmen of the country kept in subjection by the Bechuana tribe, and are a very harmless and quiet people. ...
>
> The Bechuanas throughout South Central Africa possess waggons, and have spans of oxen and everything complete like the colonists ... [They] are far more beneficial and useful in the country than the Boers. They are outstripping them in civilisation, and if they had white skins, would be looked upon as a superior race.

The culture which descended from Mapungubwe, Mmamagwa, Great Zimbabwe and Thulamela was still flourishing along the Limpopo in the late 19th century. There were vestiges of it in the late 20th century despite all the pressures placed on it.

Photo : Miriam Vigar

The author at the entrance to the Great Enclosure, Great Zimbabwe, in 1993.

Interior of the Great Enclosure, **Great Zimbabwe**, in 1971.

Photo by the author.

Terrain of the Limpopo kingdoms. Confluence of the winter-dry Shashi and Limpopo Rivers near Mapungubwe (top), and Mmamagwa hill near the Motloutse River junction.

Photos by the author.

NINE : *TERRA DA BOA GENTE*
Western Europeans explore south of the Sahara.

Great Britain annexed the Dutch colony at the Cape of Good Hope, first as a temporary measure in 1795 to frustrate the French, and then permanently after 1806. The Cape was formally transferred to British sovereignty by the Treaty of Versailles at the end of the Napoleonic Wars in 1815. Until that time, British overseas activity had been consumed by efforts to establish colonies in North America, Caribbean islands and India. Apart from Egypt and the Barbary Coast, Africa was of little interest.

The Cape, a spoil of war, was the first properly administered British colony in Africa and the European onslaught on the sub-Saharan region entered a new phase. The impact of Industrial Civilisation now became overwhelming. It may seem to have been slow at first, but in terms of our fourth dimension of time in Africa, it was lightning fast.

The southwest corner of Europe is one of those geographical crossroads where many different symbolic events occurred. It is a historical vortex. Some of the events were acts of war and some were the result of war or presaged wars and other social conflict. Christopher Columbus sailed from there towards the Americas a half-millennium ago and, as the popular media endlessly reminded us in 1992, that event resulted in the greatest colonisation by an alien race of territory occupied by another.

One of the more spectacular naval battles in this vortex was the one in which Horatio Nelson sprung into prominence as a bold and unconventional fleet tactician. This was the Battle of Cape St.Vincent on the 14th February 1797. The British fleet under Admiral Sir John Jervis consisting of 15 ships-of-the-line and five frigates outgunned and outmanoeuvred 27 Spanish ships-of-the-line with attendant frigates, which included seven giant four-deckers with more guns than any comparable British vessel at sea. The Spanish flag commander, Admiral Don José de Cordova, had the task of taking his fleet to Brest to link up with the French and Dutch in preparation for an invasion of England. The successful interception of this fleet was therefore an important strategic objective affecting the future of Britain, and therefore of Europe and the rest of the world.

Historians can endlessly debate the pivotal role of this or that particular event. I have long seen the Battle of St.Vincent as having special importance for a number of reasons, not least for launching Nelson into prominence as something more that just another good and brave naval officer. St.Vincent was the beginning of a remarkable chain of naval battles, in which Nelson dominated several, which pulled British mastery of the oceans together. In succession, other great maritime nations of Europe were beaten at sea: Spain, the Netherlands, Denmark and, most spectacularly, the French. These maritime nations had been extending the impact of European Industrial Civilisation through trade and colonisation over the world, and Britain became dominant for a hundred years which were crucial to Africa.

Because this dominance was forged in a chain of naval battles during the first global war which raged from the Caribbean and North America to India, the British maritime ascendancy thereafter had to be maintained by formal annexation of overseas territorial enclaves as bases and the imposition of political control, backed by that nonpareil naval strength. The enormous territorial extent of the British Empire, the greatest the world has known, began as a strategic necessity to protect trade, not to acquire territory for settlement. The Americas absorbed almost all surplus Europeans.

The African manifestation of the new British maritime ascendancy was the formal annexation of the Dutch enclave at the Cape in 1815 and a steady acquisition of trading castles and outposts in Ghana and Nigeria. The need to safeguard trading bases, eliminate the slave trade and ensure the security of transit ports was the limit of ambition for many years and the British followed the pragmatic and superficial policies of the Arabs, Portuguese and Dutch in sub-Sahara Africa.

Later, public excitement about explorers' journeys and exponentially expanding industrial economies seeking ever-more trade gradually forced African lands, and almost incidentally the people that lived on them, into European political control and management. Competition between European powers was resolved at the Congress of Berlin in 1884 which, although much castigated these days especially for the arbitrary and destructive national boundaries imposed by it, saved Africa from being the cockpit of European colonial wars. The notable examples of the highly destructive Anglo-Boer War and the conquest of all the German African colonies in the World War of 1914-18 by South Africans, French and British show just how bad it could have been.

Although political control of the continent by European states did not seriously begin until the late nineteenth century, the psychical and economic shock to sub-Saharan Late Iron Age Africans from Industrial Age Europeans had already begun to be increasingly devastating. The sheer power and remorseless progress of this cultural invasion had never been experienced before by the core-people of humanity.

Whenever the sub-Saharan African scene is viewed dispassionately and without sentimental or political bias, it has to be remembered that the people of Eurasia and the Mediterranean Basin diverged away from their African genetic heritage 80,000 years ago, and enormous cultural change and adaptations had occurred to both racial groups since then.

*   *

The first tentative and halting moves toward the colonisation of sub-Sahara Africa by Europeans began at the same historical vortex at the southwest corner of Europe more than three hundred years before Horatio Nelson tipped the balance in a most strategic naval battle.

At Cape St.Vincent, a younger son of a Portuguese king dreamed of trade wealth through ocean voyaging and brought together the expertise and the men to launch Christian European exploration in those parts of Africa not already occupied by or under the hegemony of Arabs and Islam. The Portuguese knew that gold and other valuable commodities, including slaves, were reaching the markets of the Barbary Coast by trans-Sahara caravan, and since conquest of North Africa was impossible, the way to intervene was to sail around to the sources.

Professor Eric Axelson in the introduction to his book *Congo to Cape* (1973) wrote:

> It was no accident that Portugal became the first European country in modern times to explore and colonise beyond the seas. Her medieval wars of independence against Leon and Castile, and her campaign against the Moors in the Iberian peninsular, had encouraged the growth of a national spirit by the time - in the middle of the twelfth century - Portugal attained what are essentially her present frontiers. Bounded by unfriendly and often actively hostile Spanish kingdoms and Muslim principalities, Portugal was forced to look to the sea not merely for communication with the rest of Christendom, but also for essential trade ...

Portugal's land is poor and there are no mineral riches. Her fishermen and sailors were tough and experienced and the exploitation of seatrading was the most natural path to expansion, if not survival, as a nation. In 1317, a friendly relationship with Genoa, Venice's chief rival for the Mediterranean trade with the Middle-East, was established when the Portuguese king appointed a Genoese as his admiral with the task of building a navy. Whereas Venice had arrangements with the Turks and the Levantines, Genoa had concluded a treaty with Egypt in 1290. Portugal, therefore, had some access to the knowledge and expertise of one of the major Mediterranean naval powers of the day and whatever secrets Genoa possessed about ship construction, Arab geographical know-how and the eastern trade.

Over the next century, Moorish naval fleets and corsairs were swept from the southern Algarve coast and Portuguese ships and sailors were tried out in an expedition to the Canary Islands. In 1415, King João I took the next step, an attack on the African mainland of Morocco, and captured Ceuta. There was some crusading zeal in this attack but there were also practical objectives: the continuing war against pirates and corsairs who menaced the Straits of Gibraltar, and because Ceuta was an important entrepôt for North African trade. Gold, ivory, slaves, hides and skins were brought to Tunis, Algiers, Tangier and Ceuta from the Sahel trading cities of West Africa.

Dom Henriques, the Infante, went to the battle for Ceuta with his father, King João I, and stayed on for a while as governor. There he learned the trans-Sahara trade and of the existence of gold in West Africa. Hanging onto Ceuta was difficult but manageable, conquering all Morocco was an impossibility. Clear logic suggested that the only way to get at the West African wealth was to make use of the ocean. Prince Henry 'the Navigator', as he has become known, determined to make use of the continuing friendly relations with Genoa to extend Portugal's knowledge of seafaring crafts. And since his idea was to tackle the Atlantic, this expertise had to be extended beyond borrowed knowledge of the Mediterranean.

The practical problems of adapting spherical trigonometry to navigating by measuring altitudes of the sun and stars from the deck of heaving ships had to be mastered. Tables of altitudes for the sun had to be calculated. Assumptions about the sky in the southern hemisphere which was unknown to them had to be made. South of the equator, there is no visible pole star with which to arrive at latitude, check compasses and calculate magnetic variation. Theories and practice of cartography had to be worked from first principles to fit into the mathematics of spherical trigonometry and celestial navigation.

It was an extraordinary task and the greatest block to a complete system was the great difficulty of calculating longitude without an accurate clock. That particular problem was not solved until the trial of Harrison's chronometer as late as 1761 provided sailors with the means to easily set a position with real accuracy at sea. Not only was it difficult to obtain the ship's position before then, but charts could not be accurately drawn. Columbus' spectacular failure to identify which continent he reached in 1492 was caused by this inability to measure longitude. It is a tribute to the Portuguese, and the legacy of Prince Henry's school at Sagres, that their navigation and cartography was quite remarkably good.

The maps of Africa and the western Indian Ocean which were drawn from Portuguese data in the 16th century were finely detailed and had relatively little distortion when considering these practical problems. Indian Ocean sailors had long been practised trans-ocean navigators by the 15th century, but they could not calculate longitude either and relied on a ponderous system of star sites to establish position. Much depended on the care and dedication applied by the navigator to his dead-reckoning of the relative passage east and west along lines of latitude. When beset by storms or beating labouriously into the wind, dead-reckoning becomes notoriously difficult. Unknown currents can make a difference of more than a degree of longitude at the equator in twenty-four hours. To this day there is controversy about exactly where Columbus made his first trans-Atlantic landfall.

Local romantic legend tells that Prince Henry was attracted to Cape St.Vincent because it pointed the way into the Atlantic and by its remote and savage beauty. Nearby, sheltered by another great promontory, there was the little fishing village of Sagres, and within a larger bay further to the east there was the port of Lagos. Everything he may have wanted was there: magnificent and strategic landscape with remote mystical appeal, peace and quiet for scholarly endeavours, local sailors used to the hard life of fishermen on the ocean waters and a town of reasonable size with a sheltered harbour. It has been said that he established a residence there, probably at the base of Cape St.Vincent. Near Sagres, he built a fortified school and library for the collection of geographical and maritime information and its study. When the time came for ships to be built and sent off, they sailed from Lagos. The school and Prince Henry's possible residence were sacked by Francis Drake in 1597 and no identifiable trace remained apart from a ruined chapel and a large monumental compass.

Henry sent out the first exploring voyages in 1422 and he died in 1460. Several voyages were undertaken along the African coast and each time lessons were learned and applied to the next one. The first years showed slow progress. The coast of Saharan Africa was uninhabited, had no water and was difficult to approach because of offshore banks: useless for trade. Adverse winds and currents had to be tested carefully. Ambitious or lazy captains often employed themselves with attacks on Moroccan corsairs and looting coastal Moslem shipping rather than pursue what often seemed an impossible task.

Because of northerly winds and current on the Moroccan Atlantic coast and the difficulty of returning, Cape Bojador (26°8' N latitude) had been held with the same dread by Arabs as Cape Correntes in Mozambique. It was the proscription by Arabs of sailing beyond Cape Bojador which created a romantic myth of mystical dangers, sea monsters and boiling seas in southern oceans. Eventually Gil Eannes discovered that he could master contrary winds that had inhibited Moroccan galleys in previous centuries by trying a bold triangular route, sailing out into the ocean, finding favourable wind systems and returning via the Azores. He returned from rounding Cape Bojador in 1434 by this method and this ancient maritime barrier was overcome.

The caravel of that time, exclusively using lateen sails previously developed in the eastern Mediterranean and Red Sea, was the usual vessel used by the Portuguese. Using a caravel, Nuño Tristão reached Cape Branco (20°46'N. latitude) in 1441 where he at last found settled people, and beyond lay prosperity. Later that decade, Tristão explored the Senegal and Gambia Rivers. The Portuguese could now begin trading directly with West Africa. In Prince Henry's lifetime, trade had been established and his youthful dream realised. That dream had not been to conquer the Indian Ocean as some have supposed; it was to bypass the Moroccans and trade with West Africa. Before Henry died, the coast of Sierra Leone had been reached.

After 1460, exploration proceeded, promoted by succeeding kings as the value of the African trade proved itself and wealth began to accumulate, justifying Henry's gamble. In 1482, the first permanent European outpost in sub-Saharan Africa was established at Elmina (5°05'N.1°21'W.) on the coast of modern Ghana. The Portuguese built the fortress of São Jorge which, with later additions by the Dutch, still stands today, grandly dominating the small sheltered bay divided by the Beya River. I have stood on the ramparts of this castle and enjoyed communing with the centuries of history held in the stones beneath my feet. Eric Axelson:

> One of [King João II's] first acts was to order the construction of a fortified trading post on the Mina coast, and in January 1482 a squadron under Diogo de Azambuja anchored off the Aldeia das Duas Partes. In the dank and equatorial heat, the Portuguese landed in their rich and colourful garments of brocade and silk, and assured the local ruler, whose scanty garb of gold chains round his neck and gold beads in his hair excited his visitors' cupidity, that their king wished to trade. ...

If southern African Bantu-speaking cattle-oriented people had little interest in gold as personal jewellery, this was not the case with West African forest farmers who had been influenced by Arab and Berber trans-Saharan traders for many centuries.

In 1486 Alfonso de Aveiro led an expedition to Benin, the sophisticated kingdom to the immediate westward of the Niger delta (not the present-day state of the same name). This was the first recorded penetration by Europeans into the rainforest zone of tropical Africa. Trading relations between the Portuguese and the Obas of Benin were established, ambassadors exchanged, slaves and peppers were sent to Portugal. The peppers began a tradition in Portugal which modern tourists enjoy when they eat chicken *piri-piri* in restaurants along the Algarve, and the gift of slaves began the trickle which led to the enormously successful colonisation of the Americas.

Competitors from other maritime nations of the Mediterranean and Atlantic seaboard of Europe were quick to follow once the Portuguese had found the way and developed the navigational and shipbuilding skills. Despite whatever military activity they could mount overseas and diplomatic pressures the Portuguese king could apply at home, it was clear that Portugal could not sustain a monopoly of the West African trade. Since Portugal could not go to war with half of Europe including her oldest ally, England, there had to be another solution other than letting her hard-won trading position melt away. This was to go on to the next giant gamble: a sea-route to the Indian Ocean and Far East, to the very source of far greater trading richness, bypassing the traditional Arab and Turkish middlemen and seatraders of the Levant, Egypt, southern Arabia and the Persian Gulf.

In Egypt, Ethiopia and the southern Sudan, notably the Kingdom of Axum, there were well-established Christian states before the foundation of Islam. The Byzantine rulers of Egypt were dislodged by an Arab Islamic army in 639 AD. Axum fell under Arab invasions, causing disruption to the general region, and Christianity retreated to the highlands of Ethiopia where it became cut off from civilised commerce.

The kings of Ethiopia, hearing about the Crusades, managed to send ambassadors to Europe from time to time seeking help and an awareness of a Christian stronghold enveloped by hostile Islamic neighbours in the heart of Africa was kept alive in Europe. The myth of Prester John, a Christian king commanding great wealth amongst 'primitive' Negroes and besieged by the hated Arabs, was frequently bandied about in Europe during the 14th and 15th centuries, brought to life occasionally by travellers' tales and the infrequent appearance of Ethiopian supplicants.

The proposition that the Portuguese were lured around Africa by the dream of linking up with Prester John to form some great Christian alliance to exploit the gold of Africa has been exaggerated, but it was a genuine motive. The Old Testament story of Solomon's gold of Ophir was there to lend strength. The religious dimension added to commercial zeal, so Portuguese explorers received instructions to seek Prester John, link with him and supplant Islam by Christianity. However, it was not a simple matter for King João II and his successor Manuel to decide to progress with exploration beyond West Africa, and this was complicated by confusion and deviousness in the court as the kings struggled to assert their authority over powerful aristocratic families and their control over the orders of chivalry.

Portugal in the last part of the 15th century, not too far away in time from its struggle for independence from Islamic Moorish rule, still had strong feudal militaristic structures which increasingly dominated commerce and the economy. The Orders of Santiago, Avis and of Christ, with their rigid hierarchies and disciplines vied with the monarchy for power, and their influences on decisions regarding exploration and trade were substantial.

The ambition to increase wealth and power through trade was strong in all parties, but precisely what strategy to follow and who was to be most closely involved caused loss of direction from time to time. João II and then Manuel (after 1495) had to decide whether the correct strategy was to play safe by consolidating Atlantic trading established during the time of Prince Henry, endeavour to obtain a share of the eastern Mediterranean trade with diplomacy and force where necessary, or find an oceanic route around Africa. The possibility of an alliance with a Christian kingdom in the heartland of Islam may have been the factor tipping the final decision in the direction of oceanic exploration.

The decision to proceed with oceanic exploration with an ultimate objective of controlling Indian Ocean trade may have been made, but the risks were well-understood. Complex and complicated politics at court and the ever-present rivalry of the

Spanish kingdoms demanded sensible caution and careful secrecy. Until the sea route to India was a proven economic possibility, all expeditions were small and economic. If three or four ships manned by a couple of hundred men were lost, it was not too great a disaster in terms of manpower, financial cost and the king's reputation.

*

Diogo Cão, a veteran of West African trading, was despatched on the first voyage of the second phase of exploration, with a clear Indian Ocean objective, in August 1482, nearly ten years before Columbus' first voyage. Having determined on the African route to the Orient, the petitioning of the Portuguese king by Columbus, the little-known Genoese trader who had sailed on Portuguese ships to West Africa, to try a westward approach was dismissed as a fantasy. The Portuguese authorities had a more accurate appreciation of the distances involved than that presented by Columbus' erroneous calculations.

As Diogo Cão, and then Bartolomeo Dias, achieved success, Columbus' desperate activity to get a sponsor to try his westward route increased. Spanish authorities would not have been ignorant of the geography and it may be that the Spanish monarchs' agreement to sponsor Columbus was prompted by little more than curiosity and indulgent patronage, tempered with greed at what he might actually achieve despite all the evidence against him. Perhaps Columbus was merely a pawn in the game between the Spanish and the Portuguese. It was a murky period in Iberian politics. It has been suggested that the Portuguese king actually mislead Columbus so that he would set off in the wrong direction, believing that his loyalty was suspect.

It is a measure of accumulating Portuguese maritime expertise and confidence that whereas it took sixty years to explore West Africa and establish a trading fort at Elmina, involving many men and different expeditions, it took six years and three recorded voyages to extend knowledge from the Bight of Benin to the Indian Ocean and one more to navigate the eastern African coast and reach India. When Columbus was exploring Central America in 1502-4, still believing that he was on the coast of Asia, the Portuguese were established on the Malabar coast of India. They achieved trading links with Canton in China eight years before the Magellan expedition returned to Europe with proof that the vast Pacific separated the Americas from Asia.

On his first expedition, Diogo Cão entered the estuary of the Congo River, went on to rest in the harbour of Luanda in Angola and reached a point of land which was later called Cabo de Santa Maria. This place lies about halfway between the modern Angolan towns of Benguela and Namibe at 13°24' S latitude. A stone pillar, a *padrão*, was erected there on 28th August 1483, and he turned back, having exhausted his provisions and the stamina of his crew.

Diogo Cão was at sea again in 1485, leaving Portugal at the end of summer, and he and his men must have been full of the possibility of sailing the Indian Ocean. He stopped in the mouth of the Congo River and exchanged friendly messages with the king of the baKongo who later became clients of Portugal. On rocks beside the falls of Yellala, the oldest surviving European graffiti in the southern hemisphere were chiselled, as translated: *"Here reached the vessels of the distinguished King Dom João II of Portugal."* Diogo Cão's name follows with some of his officers. They went on overland to pay a courtesy call on the king of the baKongo whose capital was at the modern town of São Salvador (renamed Mbanza Kongo after Angolan independence), where a Jesuit mission was eventually established. There was a faint hope that the great Congo river might provide a shortcut to the Indian ocean, but intelligence received from the baKongo dashed this.

Diogo Cão sailed on past the furthest point of his previous expedition. The coast continued southwards and shortly his ships were skirting the Namib Desert, along the notoriously dangerous land which has become known as the Skeleton Coast. He erected his last stone marker on that coast in Namibia, at Cape Cross (21°46'S), where he died.

A new voyage was immediately planned and a professional ships' captain with West African experience, Bartolomeo Dias, was given command. He made landfall and anchored for repairs and water south of Cape Cross within a beautiful rocky bay at Angra Pequena (26°38'S), where the lonely desert-begird lobster-fishing town of Lüderitz stands today. Dias' little caravels were hurtled around the Cape of Good Hope without sighting land in a storm lasting days and he did not know he was in the Indian Ocean until he found that the coast was finally heading east and north. Dias ended his voyage well into unknown seas, his ships battered and his crew exhausted, probably 150-200 kilometres past the modern city of Port Elizabeth. Apart from meeting Khoi who were not prepared to trade their cattle, the land was forbidding and apparently empty. After turning back, Dias erected a *padrão* on a rocky promontory called Kwaaihoek at 33°48'S. 26°38'E. It is from 26°28'E longitude at Cape Padrone that the coastline consistently bears away to the ENE and his mark at Kwaaihoek was chosen with care, showing the way to the Indies.

Coincident to Dias' voyage, two spies were sent out to bring back detailed evidence directly from travel in the Indian Ocean. They were to supplement hearsay and traveller's tales with firsthand knowledge of the Indian Ocean navigation systems, discover who were the important potentates, especially those who might favour trade with Portugal, and to seek substantive information about the legend of Prester John.

Pedro de Covilhão and Alfonso de Paiva were Portuguese who were familiar with Arabic and had extensive trading experience. Covilhão's instructions were to report on the Indian Ocean and Paiva was to find Prester John. Paiva did not survive to report back, but Covilhão spent three years travelling between Arabia, India, East Africa and Egypt. He sent a report back to Lisbon via Jewish traders in Cairo and then went on to follow Paiva to Ethiopia where he lived for the rest of life.

Covilhão's information was supported by Dias' voyage and by 1491 or 92 the Portuguese had positive and clear intelligence of the geographical outlines of the Indian Ocean, the commercial possibilities with experienced and sophisticated Indian princes and the political structures about it. There was no possibility that Columbus' concepts could have been accepted in Lisbon.

A summary fragment of Covilhão's report (1490) is quoted in writing by Francisco Alvarez in 1540:

> ... [Covilhão informed] the King about all he had seen along the coast of Calicut, and Hormus, and the coast of Ethiopia and Sofala, and the big island [Madagascar], saying finally that if his caravels, which were accustomed to sail Guinea, went on navigating along the coast and asked for the coast in the eastern ocean of Sofala or the island of the moon [Comoros] or the big island, they could easily penetrate to those eastern seas and reach the coast of Calicut, for it was sea all the way. ... And that the

Indian and Arab sailors knew very well the most southerly point of Africa.

It was in 1495 that King João II died and Manuel succeeded him, although there was a strong movement to have Dom Jorge succeed. Jorge was the illegitimate son of João II who had named him as his preferred successor. Hence it may be understood that this was a time when intellectual energy at court and that available to the new king was diverted. However, King João's Indian Ocean policy was actively processed by Manuel because it was in the vital interests of all parties to extend the trading wealth of the nation. At first, there was strong opposition by leading families to maritime connections with India on the grounds that the maintenance of the trade would be too costly for commercial profitability. Manuel was determined to proceed and succeeded in getting his council to agree to the monarchy taking responsibility, thus the king put up the cost and reaped the rewards. It is notable that at this time, 1495, there was no doubt about the ability of Portuguese ships to sail to and from India because the route was known. The argument was about financial viability.

According to accepted history, it was left to Vasco da Gama, a forceful aristocrat and military man, to complete the Portuguese voyages of exploration to India. Da Gama was a member of the Order of Santiago which promoted the succession of Dom Jorge to King João II and it might seem puzzling that King Manuel chose him to lead the expedition into the Indian Ocean.

Sanjay Subrahmanyam in his exhaustive biography, *The Career and Legend of Vasco da Gama* (1997), describes the lack of clarity in the record for the choosing of da Gama, a young man at the time. The complexities of court politics, the difficulties of the monarchical succession and the paucity of firsthand documents make it hard to establish precision. There is the possibility that although da Gama was a member of the Order with which Manuel was having difficulty, he had carried out duties successfully for his predecessor, João II, and it was finally because of personal affinity that King Manuel chose him. Additionally, if he failed, the fact that he belonged to the antagonistic Order of Santiago would be no disadvantage to the King.

Sanjay Subrahmanyam summarizes the Portuguese condition at this time and makes a point that I have long felt needed making in regard to the often simplistic remarks made by historians with reference to the exploration and establishment of trading colonies in the Indian Ocean.

> Portuguese historiography, for long dominated by a nostalgic nationalism, has tended to ignore the extent to which the elite of that country was divided at the turn of the sixteenth century on the question of overseas expansion. Neo-Marxist historians ... have for their part been excessively preoccupied with the issue of class conflict in this epoch, and hence also downplayed the problem. As for observers of Portuguese expansions in other historiographies (say, historians of early sixteenth-century India or East Africa), their tendency was obviously to treat 'the Portuguese' as a monolith, and then to observe this monolith undergo a cycle of rise, consolidation and decline. Vasco da Gama is thus no more or less to them than the embodiment of Portugal.

Vasco da Gama set sail in 1497 with a well-found fleet of four ships, specially designed for the Cape route by Bartolomeo Dias and armed with cannons to combat the Arab ships and forts reported on by Covilhão.

But there had always been some mystery about the ten years lapse in time between the departure of Dias' and da Gama's voyages, considering the rapid advances in the 1480s and the excitement of Columbus' activities.

Eric Axelson in *Portuguese in South-East Africa 1488-1600* (1973) stated that the Portuguese king was awaiting news from Covilhão before authorising the outfitting of a new expedition. This seems unlikely since Covilhão's despatches from Cairo were received in 1491 or 1492 at the latest. Axelson suggested elsewhere that delay was caused by the political problems besetting the rulers in Portugal which included the death of King João II and the accession of Manuel. This undoubtedly was a factor, but the momentum was great, knowledge pointing to great success had been received and competitive pressures from Spain were increasing after Columbus in 1492. Indeed, as C.R.Boxer has pointed out in *Four Centuries of Portuguese Expansion, 1415-1825* (1963), the Portuguese king was exercised after 1490 with negotiating to have the Treaty of Tordesillas agreed by the Pope which shared the 'new worlds' between Spain and Portugal. This Treaty was proclaimed in 1494, granting security for Portugal's eastern explorations and potential Indian Ocean empire. In 1494 the way was open for Portugal to proceed with no delay in establishing seaborne trade with India.

Therefore why would there have been a delay at this most critical time? As described above, all factions in Portugal favoured progress in one direction or another notwithstanding court intrigues and political manoeuvres. Since Dias' return and the receipt of Covilhão's report, planning and preparation must have proceeded.

Richard Hall in *Empires of the Monsoon* (1996) made fascinating proposals. He referred to the published work of the great Arabian navigator, Ibn Majid, in which it is reported that Portuguese ships foundered off the east coast of Africa some time before da Gama set sail in 1497. Hall also discussed the remarkable first part of da Gama's voyage in which he sets off from the Azores to sail down the centre of the Atlantic to turn east almost at the latitude of the Cape. This was the longest mid-ocean voyage ever undertaken by a European navigator until that time and implies particular knowledge of the sailing conditions and navigational requirements of the southern Atlantic Ocean for such an extraordinary feat. There is persistent opinion and some indirect evidence that the Portuguese had carried out exploratory voyages into the south Atlantic in order to investigate the wind system and circumvent the problems experienced by Dias in rounding the Cape of Good Hope. In 1486 two voyages of exploration westward into the Atlantic, possibly as advance planning for Dias, were authorised, but the results are unknown.

It is notable that King João II insisted that the line dividing Portuguese and Spanish spheres of influence in the Treaty of Tordesillas of 7 June 1494 should be drawn in such a way as to include the northeast coast of Brazil. The line was defined as being 370 leagues (1770 kms) west of the Cape Verde Islands at ± 46°W. Surely this was no coincidence? Axelson in *Congo to Cape* (1973) reports that Dias was appointed to the staff of the Guinea House, responsible for trade to West Africa, in 1494. He was given authority to design and oversee construction of the improved, square-rigged *naus* of 200 tons which could take advantage of mid-ocean sailing in the trade wind belts.

Hall's assumption is that there were further voyages between the return of Dias and the departure of da Gama and that records have been lost. Many Portuguese archives were destroyed by fire in the Lisbon earthquake of 1755 and it would not be surprising if this was so. It is an exciting notion. These suppositions were directly supported by Arab sources.

Ibn Majid's *Sofaliya* (c1500):

> Ahead [of Sofala] - well-taken care of ! - the wind *kus* [the southerly monsoon] cools. You go up to the sunrise, its land (is labourious), it arises close by - look at it with attention, pilot. The waves go up in these instances. It was here that the Francs [the Portuguese] stumbled, they had confidence in the monsoon, on the day of Saint Miguel [Michael], or so it seems. The waves precipitated them, in these lands of Sofala, [the ships] turning over. The masts dived in the water, and the ships had been underneath the water, O my brother. Some of them drowned - they know thus what it is the monsoon of this land.

And later, when describing the coastlands about southern Africa, the *Sofaliya* continues:

> Here the ships of Francs [Portuguese] had passed in the year 900 (1495-1496), brother. They had sailed for two entire years before arriving here, and were passing here, evidently, on the way to India. Who searches for Sin (China) watches the dangers; in all other cases he does not have hope. This way the Francs had come back from India and the Zanj and later, in the year 906, new ones had arrived in India, my brother.

The feast of Michael makes a precise date of 29 September 1495 for the disaster in the vicinity of Sofala. September-October is the time of the changing of the monsoons, explaining why a Portuguese fleet "stumbled", because they had false "confidence" in the monsoon.

I see no reason to doubt these passages from Majid's *Sofaliya*, whether he wrote, or dictated, them himself, or whether they were added later by an anonymous editor. Some commentators have cast doubt on the facts described in the epic primarily because they are not supported by Portuguese records, and a new dimension is added to the old-established history of Portuguese exploration and da Gama's voyage. However, there are no Portuguese records available for this or other presumed voyages in 1494-96 and the evidence must rely on the authenticity of the *Sofaliya's* report.

I had assumed that the Portuguese had to have carried out further reconnaissance of the Atlantic Ocean weather systems at the least in order for da Gama to have undertaken his seemingly extraordinary tactic of sailing boldly down the western side of the South Atlantic until he reached almost precisely the correct latitude to bear off east to land at St. Helena Bay at the Cape. The boldness of the move involving so many days at sea and the strange precision of his arrival at St. Helena (32°45'S), a very suitable landfall and place to refresh and repair on an inhospitable coast, seemed a carefully considered tactic based on foreknowledge and not some brave gamble.

There is also the apparently unimportant detail of da Gama deliberately keeping out to sea to bypass Sofala (the fabled source of African gold at 20°10'S. lat.) because he feared becoming caught on a lee shore during the period of the southerly monsoon. Sofala lies within a deep indentation of the coast and da Gama had already experienced the problem of a lee shore at Aguada da Boa Paz in southern Mozambique in January 1498. Clearly, he knew the navigational problems of that region, whether from personal experience or from the reports of the earlier expedition as described in the *Sofaliya*. The 1497-99 expedition to India, with all the years of preparation and risk to the kingdom, would not have been undertaken without much planning and endless discussion. It does not diminish da Gama's exploit, especially if, as has been proposed, he was himself on the voyage of 1494-96.

Eric Axelson in *Vasco Da Gama* (1998) :

> There is, however, further evidence of a possible Portuguese expedition to East Africa before 1498: the so-called Cantino map, bought by an Italian agent in Lisbon at the end of October 1502. Armando Cortesão established that none of the three fleets arriving in Lisbon before 1503 stayed long enough in African waters to make such a detailed hydrographical survey of East Africa as was shown on this map. How then to explain the near perfection of the representation of Africa in that planisphere; which implies a careful survey. ...

The implication that a careful survey was undertaken takes me to the apparently loose statement in the ibn Majid *Sofaliya*, wherein it is stated that the Portuguese had sailed "for two entire years" before reaching Sofala. That period would have been too long for a normal voyage, following in the footsteps of Dias, but would have been profitably used in mapmaking.

Apart from anything else, the very design of the ships had been subject to much thought and a new sailing rig developed which was to be the basis for all future European ocean-going ships. Dias' ships were rigged with two or three masts and only fore-and-aft lateen sails on each. This was a tried design in use in the Mediterranean for centuries and arguably was derived from the ancient Arab Indian Ocean rig used on their several classes of dhows. This lateen rig had served during coastwise exploration of the West African coast, but Dias' experiences of the contrary winds and storms off the Cape of Good Hope required urgent modification. Changing tack or wearing ship with a lateen rig with its long yard and short mast is labourious at any time and dangerous in strong winds. This had always inhibited Arab navigation into the uncertain winds of the southern Indian Ocean. No doubt, assuming there had been an exploratory voyage after Dias to determine the wind systems of the South Atlantic, more experience of mid-ocean sailing supported innovation.

The new and improved Portuguese caravels of the 1490s, instigated by Bartolomeo Dias, had square sails on the fore and main masts, easily braced around from one tack to the other, and a fore-and-aft lateen sail on the mizzen. The classic full-rigged ship and three-masted barque of the finest late 19th century vessels are sophisticated refinements of this sail plan. In addition to developing a radical new sailing rig which served all large modern oceanic sailing vessels, the new caravels were bigger and more strongly constructed with a higher freeboard to combat the stormy seas of the southern oceans.

Ibn Majid, the author of navigation manuals and descriptions of the Indian ocean, is often quoted as conducting da Gama to India from Malindi to Calicut in 1498 and it is undoubted that they met during April 1498. Certainly ibn Majid received information from da Gama himself and others of the fleet about earlier Portuguese voyages, apart from other intelligence that he would have gathered from his own Arab sources, or even from firsthand knowledge.

It is not beyond belief that he had earlier dealings with Portuguese navigators, as suggested in the *Sofaliya*. Sanjay Subrahmanyam in *The Career and Legend of Vasco da Gama* (1997) is much concerned with Portuguese historical resources in pursuing his extensive and updated biography and sees that discussion of voyages into the Indian Ocean is subordinate to a full understanding of the 'environment' of Vasco da Gama. He treats the probable earlier voyages and Armando Cortesão's arguments in *The Mystery of Vasco da Gama* (1973), this way:

> Armando Cortesão suggested that the 'mystery of Vasco da Gama' could be solved if one posited that between 1487 and the death of Dom João II in October 1495, there had been some Portuguese expeditions to the east coast of Africa round the Cape of Good Hope. Further, he proposed that one of these, in 1494-95, was led by Vasco da Gama himself, but that by the time of his return, Dom João II was already dead. Dom João's successor Dom Manuel, not wishing to give the dead king credit for the entry into the Indian Ocean, suppressed information of the expedition's success, and then himself mounted another one ...

Subrahmanyam sees this as a naive opinion which does not take into account all the complexities of the situation. I am happy that his exposition of these complexities of da Gama's 'environment' sheds far greater light on the background to Portuguese expansion into the Indian Ocean than has been easily available in English until now, in the same way that Peter Russell's biography, *Prince Henry, 'the Navigator'* (2000), has done for the beginning of this period. However, my own fascination is with the voyages themselves and the interaction between the first European seatraders with the ancient Indian Ocean trading environment.

G.R.Tibbetts, the author of *Arab Navigation in the Indian Ocean before the coming of the Portuguese* (1971), and a devoted scholar of ibn Majid's work, has little doubt about the authenticity of his *Sofaliya* and goes thoroughly into the question of the ibn Majid and da Gama association. There seems no question that the two men met in Malindi in 1498 and exchanged information on themselves and navigation during extended discussions. That both men were most interested in what the other would have to say is undoubted and da Gama would be avid for ibn Majid's knowledge. A Sunni Moslem historian from Mecca, Qu b al-Din al-Nahraw, described one of their meetings, probably with exaggerated malice since ibn Majid was a Shiite from Oman. Nevertheless, he confirmed the association by writing that ibn Majid "... became bewitched with the Portuguese admiral" and "being drunk, showed the route [to India] to the admiral ..." Qu b al-Din blames ibn Majid for the arrival of the "accursed Portuguese".

It seems perfectly reasonable to me that ibn Majid was plied with wine by da Gama and maybe they had several jolly meetings as their friendship was established and they exchanged intimacies. They were both proud men and ibn Majid is prone to boasting in his writing. What is in serious doubt is whether ibn Majid was the pilot who actually accompanied da Gama to India as is normally stated in general histories.

A pilot was engaged by da Gama in Malindi for the voyage to Calicut, but he is described in various Portuguese documents as the 'Moor from Gujerat', nicknamed *Malemo Canaqua* (roughly translated as the 'navigator-astrologer'). Ibn Majid was an Omani Arab of considerable renown and would never have been described as being an Indian from Gujerat. He met da Gama and maybe told him much about the Indian Ocean, but da Gama hired an unemployed pilot, presumably Moslem, who was from Gujerat (the Indian coast north of Bombay).

João da Barros (1496-1570), for example, wrote:

> Among them was a Moor [Moslem] of the Gujerat nation, named Malemo Cana, who from the pleasure he took in the conversation of our people and his desire to please the king [of Malindi] who was seeking a pilot for them, agreed to go with them. After da Gama had conversed with him he was well satisfied with his knowledge especially when he showed him a chart of the whole coast of India arranged in the Moorish fashion ...

Ibn Majid in his own writings describes the first Portuguese voyages in the Indian Ocean as being told to him directly, but he does not mention sailing on one of their vessels to India. He bewailed the coming of the Portuguese, but does not blame himself for it. As Tibbetts states:

> It is much more likely that the pilot of Vasco da Gama was an Indian stranded in Africa, hoping to earn his passage back to his native land than an Arab of Ibn Majid's knowledge who would have realised the consequences of introducing the cursed Franks into Indian Ocean trade.
> ... it can only have been by some man from one of the continually warring Indian states hoping to boost his own countries fortunes at the expense of a neighbour.

Sanjay Subrahmanyam in *The Career and Legend of Vasco da Gama* (1997) has no doubt that da Gama's pilot from Malindi to Calicut was not ibn Majid. Indeed, he asserts from his great depth of modern research that the pilot remained with the flotilla and da Gama brought him back to Portugal in 1499. Subrahmanyam:

> The last nail in the 'Ibn Majid hypothesis' [that he piloted da Gama to Calicut] was driven in by the discovery of some Italian letters written from Lisbon to Florence, immediately after Gama's return. Here, there is mention of how Gama had brought back the Moorish pilot (who apparently could speak some Italian!) with him to Lisbon, where the Italian representatives of the Sernigi-Marchionni combine were able to question him.

It must also be emphasised that there were four Indian vessels in Malindi when da Gama's fleet was there in 1498. The Indians were several times described as Christian in the *Roteiro* of da Gama's voyage and the appearance and behaviour of the men themselves were described in detail. Since the *Roteiro* was completed after the journey and sojourn in India, it is unlikely that the author would have been confused about the identity of these seatraders (as has been frequently suggested by historiographers). There have been Christian Indian seafarers since St.Thomas' time and most ocean-fishing communities on the Malabar Coast today are still Christian. It is often overlooked that Christianity thrives in western India on the coast and has been there far longer than in western Europe.

Vasco da Gama's fleet arrived at Calicut, the preferred choice for Portuguese contact, as already established by Pedro de Covilhão, on 20 May 1498 and was welcomed by the Zamorin ruler. He could not conclude a treaty because of the powerful opposition of local Moslem merchants, but the way was open.

*   *

Notice should be taken of Chinese activity in the Indian Ocean before the arrival of the Portuguese. In the earlier chapter on seatraders I briefly described the voyages under the command of the great Admiral Zheng He. The Chinese fleets were quite extraordinary for their time: in size and sophistication of ship design, number of vessels, the care and planning devoted to the voyages and what was required of the complements of navigators, scientists, traders and recorders. There is no question that these exploration expeditions were the most important and the most comprehensive launched by any nation since the Romans until the eighteenth and nineteenth centuries. That the Chinese emperor decided under pressure from his conservative Mandarin bureaucrats to cease this activity in 1435 is a most remarkable turn in the history of the Indian Ocean, and Africa. Probably, it is one of the most important decisions made in modern world history. Who knows what path events would have taken had the Chinese decided to proceed with creating a trading hegemony over the Indian Ocean, and maybe further?

In 2002 a remarkable book was published by Gavin Menzies: *1421 The Year that China Discovered the World.* This book, which claimed support from various maps and documents describes Chinese voyages all around the globe with settlements or landings in several places in the Antipodes and the Atlantic shores of the Americas. This book created something of a furore but it may be discounted as yet another well-researched fictional blockbuster masquerading as history. It would have been extraordinary if the Chinese expeditions seeking new trade-links had failed to investigate and make contact with Europeans who were the potential third force, after the Arabs and Indians, in world trade and industry. The Chinese were well aware of European nations and the geography of Europe from the centuries-long trade based on the 'Silk Road' of central Asia. Indeed, the principle objective of the maritime expeditions was to establish alternative trading corridors, substituting for the 'Silk Road' which was menaced by the Mongols and hostile tribes.

The Chinese fleets under Zhen He explored the Indian Ocean and made intimate contact with Africans, Arabs and Indians, but went no further.

*   *

En route to India in 1497, Vasco da Gama's flotilla met Khoi at Mossel Bay on the Indian Ocean coast of South Africa, where he and his men noted their fine cattle and danced happily in unison with the herdsmen, the Khoi playing music on flutes and the Portuguese playing their trumpets. There were no Bantu-speaking Negro people in that part of Africa, and it was not until they made a landfall considerably further north that they met them.

The flotilla anchored offshore at the *Aguada da Boa Paz* - Watering Place of Good Peace (24°52'S. 34°25'E.) - south of Cape Correntes, on 10th/11th January 1498 and there was an amicable meeting with a large Bantu-speaking clan led by a chief who welcomed them hospitably. There was an exchange of gifts and the chief had his people assist with watering the ships. It was noted that the people wore much copper jewellery and had wrought-iron weapons, which they would have traded with upcountry clans. There was no mention of cattle. Da Gama was so impressed by the reception he met from the people of this coast that he named the land: *Terra da Boa Gente* - The Land of Good People. The Portuguese retained that delightful and historic name for Inhambane District until Mozambique became independent nearly 500 years later when it was discarded for being 'colonialist'.

The meeting at *Aguada da Boa Paz*, in the *Terra da Boa Gente,* may seem trivial and unimportant, but I have always seen it as a symbol of Portuguese intentions and first acts on the Indian Ocean shores of Africa. The Portuguese never set out to conquer territory in Africa: their objective was always trade, and trade cannot flourish within conflict.

It was the struggle for commercial supremacy on the high seas between Portuguese and Indian Ocean Islamic seatraders that resulted in bloodshed and the sacking of port-towns by one side then the another. Native Africans were the innocent victims in these skirmishes and violent disputes and were never the principals nor the cause. Portuguese commanders laid waste to Kilwa when the Sultan was intransigent early in the 16th century, but this was a relatively minor incident compared to the terrible siege of Fort Jesus in Mombasa by the Omani Arabs, lasting from March 1696 to December 1698. Mombasa was several times raided and sacked by both Portuguese and Arabs.

During the 16th and 17th centuries, ill-fated Portuguese expeditions were sent up the Zambezi seeking control of the dwindling gold production and weak military intervention was mounted from time to time to try and support native trading partners who were propped up as kings of Monomotapa in north-eastern Zimbabwe. But these were not effectively pursued. On the Angolan coast of western Africa in the 18th and 19th centuries, slaving activity for the trade to the Brazilian sugar plantations was substantial and, ironically, it was this concentration of Portuguese colonial effort in Brazil that kept them out of interior Africa. Always, the Americas were the magnet for European settlement.

Portugal established a loose and inefficient hegemony over the seatrading systems of Swahilis, Arabs, Persians, Turks and Indians for no more than a hundred years. After 1600, it was the Dutch and the British, and the French, who gradually became masters of the Indian Ocean. The comparison between the relatively benign activities of the Portuguese in eastern Africa compared to Spanish, French, British and their own behaviour in the Americas and Caribbean islands is not always known or understood.

In 1971, in unpublished papers, I wrote:

> Where real martial strength was required by the Portuguese in the Indian Ocean it was in defence against the natural reaction of the Arabs to their arrival in the Arabian Sea and in their holding out against other Moslems of Egypt, Persia and northern India who rallied to the cause. The 'White Moors' of the north [the people of Arabia] could not stand idle and see the Portuguese, in one stroke, break their monopoly of intercontinental trade with India and the East which was being funnelled through them to the Italians, and cut off entirely their trade with Africa.
>
> The Portuguese established friendly relations with a selection of Hindu princes on the Malabar Coast of India

and trade began. But the Moslem ruler of Diu could not tolerate the Portuguese. He defied them and his forces were defeated so he sent out a call for help which was answered by the Turkish Sultan of Egypt. Fleets and armies were gathered together and in the first sweep south a Portuguese squadron was overcome. The fate of the Indian Ocean was in balance.

Francisco d'Almeida, one of the more famous Portuguese colonial figures, rallied his forces and in a great naval battle of 2nd February 1509, off Diu, defeated the combined Moslem fleets. For a hundred years, despite repeated efforts by Islamic nations, and several wars in the Arabian Sea zone, the Portuguese were the masters of the ocean from the Cape of Good Hope to the East Indies. In the end it was the Dutch and English East India Companies which broke this monopoly in the 17th century by sailing directly across the Indian Ocean, bypassing the eastern African coast, with improved ships and better navigational technique. They established refreshment bases at St. Helena Island, the Cape and Mauritius and later called on the coast of Madagascar.

Ahmed Hamoud al-Maamiry was born in Zanzibar of Omani descent and has written several books on Omani and East African history, compiling material from sources not always consulted by European or African writers. In *Omani-Portuguese History* (1982) he states:

The Portuguese have been described as people who had no colonial interests in their adventures and they did not interfere in the administration of the local affairs of the countries they visited. Their principal aim was to monopolise trade from the Arabs and to convert people to Christianity. Their treatment of the local people depended upon the response they received from them on their arrival. If they were resented they used force, and if they were received with respect they reciprocated, but this treatment depended upon the attitude of the man in command. There are references of use of force even before the behaviour of the local people was ascertained.

With the failing of the gold trade from Zimbabwe, the Portuguese presence in Mozambique and eastern Africa was never profitable. Until the 19th century, their only achievements were to temporarily disrupt the trade with India, Arabia and the Persian Gulf and generate resentment from those colonial Arab and Swahili rulers and merchants who lost their lucrative dominance.

A small stone fort was built at Sofala in 1505. A fine fortress in classical renaissance style later rose in fits and starts on Moçambique Island but it was to counter expected rival European aggression. This fortress withstood three terrible sieges by the Dutch and forced them to find other routes to the Indies. It became the headquarters of the Portuguese presence on the African coast and is without doubt the finest building of its type in sub-Saharan Africa. In East Africa, the massive if ugly monster of Fort Jesus was built hastily at Mombasa when Omani strength was growing. There was no extensive building programme as there was in India.

What the Portuguese were defending was not an African colonial empire but their sea route to India. Their disputes were with Omani and Arab-Swahili dynastic and feudal city-governments over control of trading depots. Their futile expeditions on the Zambezi were to support warring local tribal chiefs who they saw as friends and trading partners. Some late twentieth century writers have relentlessly portrayed the Portuguese as colonialist plunderers of eastern Africa but the facts show this to be fashionable rhetoric and often racist propaganda.

The same general principle was applied by the other major colonising nations; the Dutch, the French and the British. As Niall Ferguson has said, succinctly, in his exhaustive book, *Empire - How Britain made the modern world* (2003), when summarizing the eighteenth century: "The British had been attracted to Asia by trade. They had been attracted to America by land."

Africa was not a factor at all. Clive began the conquest of Indian territory after he defeated French aspirations at the historic Battle of Plassey in 1757. Although Plassey may be said to be the beginning of British occupation of India, starting with Bengal, and hence the beginning of the territorial expansion of the British Empire outside North America and the Caribbean, it may also be said that it happened mainly as a reaction to French trading activity. France was beginning to gain a portion of the Indian textile trade unacceptable to British merchants and traders.

Until the Battle of Trafalgar in 1805 finally doomed French oceanic expansion, the two old enemies fought for dominance and it was this rivalry which directly impacted on the poor British performance both militarily and diplomatically in North America and lost her the North American colonies which formed the United States.

Conflict between the competing European ocean trading nations had negligible direct effects on the eastern African coast in the 17th and 18th centuries and the people of the interior were unaffected away from the old-established Swahili towns. Indeed, the Omani Arabs were considerably more successful in establishing a colonial presence in East Africa, ruling from their bases at Mombasa and Zanzibar until the British took them under 'protection' in the latter part of the 19th century.

Omani Arab merchant families established control over the major trading towns of Pate, Malindi, Mombasa and Zanzibar after expelling the Portuguese in the early 18th century but they warred with each other interminably. It was the empire-building Sayyid Said, Sultan of Oman, who unified Arab control of the East African Swahili coast after defeating the Masrui family of Mombasa in 1828. He decided to transfer his capital to Zanzibar in 1832 and ruled Oman from that base. Sayyid Said encouraged a new initiative to exploit trade in ivory and slaves in the interior, and inaugurated an efficient slave-plantation industry on Zanzibar and Pemba, along the Kenya mainland coast and in the Lamu archipelago. Arab and Swahili traders penetrated to lakes Victoria, Malawi and Tanganyika and worked along the upper reaches of the Congo River. Swahili became a *lingua franca* in much of this region of eastern-central Africa.

The writings of Burton, Speke, Stanley and Livingstone describe the enormous extent of Arab-Swahili influence. I met a sheik in Suna, near Lake Victoria, in 1985 who could remember his grandfather telling tales of his slave-trading exploits.

Eric Axelson sums up the first European colonial activity in eastern and southern Africa quite neatly in *Portuguese in Southern Africa 1488-1600* (1973):

> By and large, however, Portugal had made a notable contribution to Europe of the physical and human geography of a large expanse of Africa. But of course Portugal had not come to south-east Africa to indulge in individual's scientific curiosity. She had come primarily for strategic and commercial reasons.

> It was the maritime traffic to and - especially - from India that was of paramount importance to the Portuguese monarchs and their advisers throughout the sixteenth century. In this light the continent's southerly projection was only an unfortunate impediment that lengthened, complicated and prejudiced the voyage. Moçambique [Island] was established primarily as a marine station ...

Equally, the Dutch did not come to South Africa to establish a settler-colony in the 17th century. Their prize was always trade with the East Indies and China. When the Dutch founded the great city of Cape Town in 1652, they endeavoured for the next 140 years to restrict settlement and contain the energies of the few numbers of Dutch, French and German townsmen, peasant farmers and ranchers whose descendants became the 'white' Afrikaners of today. Cape Town was established as a refreshment station on the Dutch route to the Indies which became a necessity after their failure to wrest Moçambique Island from the Portuguese.

The ownership of the Cape was one of several reasons for the reopening of the worldwide Napoleonic Wars between Britain and her allies and France in 1803 although the trigger was control of Malta. With Britain in the ascendancy in India, an African base on the way was imperative. The Cape of Good Hope was a reluctant colonial outpost which had to be brought under British control, an essential cog in the vastly more important machine tapping into the wealth of the Indian Ocean trading system.

Livingstone's writings and public lectures undoubtedly stimulated the exploration and then colonisation of the interior of Africa. There were a number of other explorers from the several nations in Europe who played a similar role, and particular British, French, Italian, Austrian and German names are written into the history and geography of Africa. But, Livingstone was arguably the most potent publicist of African colonisation. His theme was that systematic British colonisation and orderly commercial development was needed to counter the increasingly widespread Arab and Swahili slave-trade and 'civilise' the natives.

He believed that colonisation from one source or another was inevitable and that Victorian British ethics would be the best safeguard for indigenous Africans.

The engine of the modern Industrial Revolution was running too fast for there to be any brake on colonial activity by Europeans, no matter how much ethical and enlightened arguments were also stirred up by the manner of it. There is no doubt that the European colonisation of Africa in the 19th century was inevitable. It was an historical and evolutionary imperative: as certain as night following day.

A current perspective which I believe to be of importance is Professor Niall Ferguson's book. His work describes and explains the motives and progress of British empire-building, the greatest the world has known, with all its good and bad effects. In his conclusion he states what I have always seen to be the central burden of the European colonial period. He is referring to the British Empire, but his remarks could equally apply to the Portuguese and the French in Africa.

> Though its imperialism was not wholly absent-minded, Britain did not set out to rule a quarter of the world's land surface. As seen, its empire began as a network of coastal bases and informal spheres of influence, much like the post-1945 American 'empire'. But real and perceived threats to their commercial interests constantly tempted the British to progress from informal to formal imperialism. That is how so much of the atlas came to be coloured red.

Discussion will proceed, for a long time to come, about the methods and the results.

But what if the colonisation of Africa had been delayed until the 1930s. It was at that time that Italy consolidated its hold on Libya and conquered Ethiopia, Somaliland and Eritrea. How would it have been if Africa had been subject to the regimes of Nazi Germany, Soviet Russia or to Imperial Japan?

As Niall Ferguson wrote, if Africa was at all misused in its colonisation by the British, then it received great benefit by being defended by the British Empire, with its Victorian morals and ethics born of the Enlightenment, in the mid-20th century. It is ironic that Britain, by becoming deeply indebted to the United States, lost its empire in the noble cause of defending all its colonial peoples from the evils of racist Fascism.

O mar salgado, quanto do teu sal
São lágrimas de Portugal!
Por te cruzarmos, quantas mães choraram,
Quantos filhos em vão rezaram!
Quantas noivas ficarem por casar
Para que fosses nosso, O mar!

"O salty sea, how much of your salt
Are tears of Portugal!
For us to sail you, how many mothers cried,
How many sons prayed in vain!
How many brides forewent marriage
So you could be ours, O sea!"

Fernando Pessoa (1855 -1935).

For to admire an' for to see,
For to be'old this world so wide -
It never done no good to me,
But I can't drop it if I tried.

Rudyard Kipling (1865 -1936).

Don't preserve my customs
As some fine curios
To suit some white historian's tastes.
There's nothing artificial
That beats the natural way
In culture and ideals of life.
Let me play with the whiteman's ways
Let me work with the blackman's brains
Let my affairs sort themselves out.
Then in sweet rebirth
I'll rise a better man
Not ashamed to face the world.

Dennis Chukude Osadebay (1911-1995)

The castle at Elmina in Ghana, 1995. Founded by the Portuguese in 1482.

The chapel of 'Our Lady of the Bulwark' below the walls of the
16th century fortress on Moçambique Island in 1971.

Photos by the author.

BIBLIOGRAPHY and further reading

This bibliography was compiled to go with my original publication of *Seashore Man & African Eve* and may seem to be inappropriate for this book. However, it is included here because it not only acts as a reference to the works which I have quoted in the text, but may also assist in tracing some papers and books which are not so well known in a general study of Africa.

Abysov, S.S. et al : *Deciphering Mysteries of Past Climate from Antarctic Ice Cores.* American Geophysical Union - *Earth in Space*, v.8 no.3. 1995.
Adams, W.M.: *Definition and Development in African Indigenous Irrigation.* Azania, 1989
Al-Maamiry, Ahmed Hamoud: *Oman and East Africa.* Lancers Books, 1979
   - - : *Omani-Portuguese History.* Lancers Publishers, 1982
Allen, Anita: *Exciting Hominid Fossils Discovered.* The Star, Johannesburg, 29 October 1993
  - - : *Royal Graves Found in Kruger.* The Star, Johannesburg, 7 August 1996
  - - : *Fossil Find takes Anthropological Theory into the Desert.* The Star, Johannesburg, 12 Aug, 1996
  - - : *Stone-walled Ruins with a Unique 'Sense of Lightness'.* The Star, Johannesburg, 24 Sep, 1996
Ambrose, Stanley H. : *Late Pleistocene Human Population Bottlenecks Volcanic Winter, and Differentiation of Modern Humans.* Extract from : *Journey of Human Evolution*, Bradshaw Foundation, 1998.
Andaman Association (Ed. Weber, George) : *Lonely Islands*, Internet (2006)
Anderson, Andrew A: *Twenty-Five Years in a Waggon.* Chapman & Hall, 1888, (Struik, 1974)
Anderson, David M.: *Agriculture and Irrigation Technology at Lake Baringo*, Azania, 1989
Anderson, Gavin : *Bushman Rock Art, South Africa.* Art Publishers, 2004
Anderson, Julie R. & Ahmed, Salah Mahomed. *Revealing terra incognita: Dangeil, Sudan.* Current World Archaeology no. 19, October 2006.
Ardrey, Robert: *African Genesis.* Collins, 1961
  - - : *The Territorial Imperative.* Collins, 1967
  - - : *The Hunting Hypothesis.* Collins, 1976
Asher, Michael : *In Search of the Forty Days Road.* Longmans, 1984.
Athena Review : *Levant Suite Yields Oldest Evidence of Cereal Processing.* v 4 (2). 2005
Athena Review : *Oldest* Homo sapiens *Evidence at Omo, Ethiopia.* V. 4 (2). 2005
Axelson, Eric: *South-East Africa, 1488-1530.* Longmans, 1940
  - - : *Congo to Cape.* Faber & Faber, 1973
  - - : *Portuguese in South-East Africa 1488-1600.* Struik, 1973
  - - : *Vasco da Gama.* Stephan Phillips, 1998
Baker, Samuel White : *The Albert N'Yanza.* Sidgwick & Jackson, 1962 (1868)
Barker, Graeme: *Economic Models for the Manekweni Zimbabwe, Mozambique.* Azania XIII, 1978
Barrow, John & Tipler, Frank J.: *The Anthropic Cosmological Principle.* Oxford, 1986
Battuta, Ibn: *Travels in Asia and Africa 1325-1354.* Routledge & Kegan Paul,1929
Beach, D.N.: *The Shona and Zimbabwe 900-1850.* Mambo Press, 1980
Beaumont, P.B. & Boshier, Adrian: *Some comments on Recent Findings at Border Cave, Northern Natal.* South African Journal of Science, 1972
Behar, Doran M., et al : *The Dawn of Human Matrilineal Diversity.* The American Journal of Human Genetics. 82/5, 24 April 2008.
Bent, J.Theodore : *The Ruined Cities of Mashonaland..* Longmans, Green, 1896
Bergerot, Sylvie & Robert, Eric: *Namib, Dawn to Twilight.* Southern Books, 1989
Berry, Adrian: *The Ozone Layer Was Destroyed Before.* The Daily Telegraph, London, 21 December 1991
  - - : *2001 Replayed.* Astronomy Now, March 2006

Bieber, Heidi; Bieber, Sebastian W; Rodewald, Alexander & Christiansen, Kerrin *Genetic study of African populations: Polymorphisms of the plasma proteins TF, PI, FI3B, and AHSG in populations of Namibia and Mozambique* (Human Biology, Feb 1997).
Bittner, Maximillian, Ed.: *Das Indischen Seespiegels*. Geographischen Gesellschaft, Vienna, 1897
Blench, Roger. *New paleozoogeographical evidence for the settlement of Madagascar*. Azania XLII, 2007
Böeseken, A.J.: *Slaves and Free Blacks at the Cape 1658-1700*. Tafelberg,1977
Boonzaier, Emile; Berens, Penny; Malherbe, Candy; Smith, Andy : *The Cape Herders*. David Philip & Ohio University Press, 1996
Bosman, Paul & Hall-Martin : Anthony: *Elephants of Africa*. Struikhof, 1986
Bosman-Nasmith, Ria: *My Ryperd, my Man en Ek*. Heilbron Herald, 1973
Bourne, Harry : *The Phoenicians in East Africa*. WWW. 2003
Bousman, C.Britt. *The Chronological Evidence for the Introduction of Domestic Stock into Southern Africa* . African Archaeological Review, 1998
Boxer, C.R. : *Four Centuries of Portuguese Expansion*. . Witwatersrand University Press. 1963
Boxer. C.R. : *The Tragic History of the Sea 1559-1565*. Hakluyt Society & Cam,bridge University Press. 1968
Brace, C..Loring: *Background for the Peopling of the New World*. Athena Review, 2002
Brandt, Willy (Intro.): The Independent Commission on International Development Issues: *North-South*. Pan, 1980
Brunet, Michel et al : *A new hominid from the Upper Miocene of Chad, Central Africa* .Nature 418, 2002
Broadhurst, C.Leigh et al. *Brain-specific Lipids from Marine, Lacustrine, or terrestrial food resources: potential impact on early African Homo Sapiens*. CBP, 2002
Broadhurst, C.Leigh ; Cunnane, Stephen C.; Crawford, Michael A. : *Rift valley lake fish and shellfish provided brain-specific nutrition for early Homo*. Nutrition Society 1998.
Brook, Ed. *Windows on the Greenhouse*. Nature, 15 May 2008
Brooks, Nic; di Lernia, Savino; Drake, Nic; Raffin, Margaret; Savage, Tony. *The Geoarchaeology of Western Sahara*. Draft paper for *Sahara*, 2003. www.cru.uea.ac.uk
Brown, Mervyn: *Madagascar Rediscovered*. Tunnacliffe, 1978
Cachel, Susan: *The Paleobiology of Homo Erectus and early Hominid Dispersal*. Athena Review. 2004
Campbell, John: *Travels in South Africa*, Black, Parry & Co, 1815
Carr, M.J. & A.C. and Jacobson, L.: *Hut Remains and Related Features from the Zerrissene*. Cimbebasia, 2-11, Feb. 1978
Casson, Lionel : *The Ancient Mariners*. Victor Gollancz, 1959
-         : *The Periplus Maris Erythraei*. Princeton University Press. 1989
Chami, Felix : *The First Millennium AD on the East Coast*. BIEA 1996
-    -   : *The Early Iron Age on Mafia Island and its relationship with the Mainland*. Azania   XXXIV 1999
-    -   : *The Graeco-Romans and Paanchea/Azania : sailing in the Erythraean Sea*. British Museum 2002
-    - & Kwekason, Amandus : *Neolithic Pottery traditions from the Islands, the Coast and the Interior of East Africa*. African Archaeological Review , 2003
-    -   : *Kaole and the Swahili World*. Extract from *Studies in the African Past 2*. University of Dar-es-Salaam, 2003
-    -   : *The Unity of African Ancient History 3000 BC to AD 500*. E & D , Dar-es-Salaam, 2006
Carbonell, Eudald, et al : *The First Hominin in Europe*. Nature, vol 452. March 2008
Clowes, Wm.Laird (Ed.): *The Royal Navy*. Sampson Low, Marston, 1897
Collins, Andrew. : *Gods of Eden*. Headline, 1998
-       -   : *In defence of Cygnus X-3 as a Cosmic Accelerator*. Unpublished paper, 2006
-       -   : *The Cygnus Mystery*. Watkins Publishing , 2006
Connor, Steve : *DNA Tests trace Adam to Africa*. London Sunday Times, 9-11-1997
Cooke, H.B.S.: *Preservation of the Sterkfontein Ape-man Cave Site, South Africa* . Studies in Speleology 2-1, 1969
Cordova, France Anne-Dominic : *Cygnus X-3 and the Case for Simultaneous Multifrequency Observations*. Los Alamos Science, Spring 1986
Crawford, Michael & Crawford, Sheilagh : *What We Eat Today*. Neville Spearman, 1972
Crawford, Michael & Marsh, David: *The Driving Force*. Heinemann, 1989
Crawford, Michael A., et al : *Brain-specific Lipids from Marine, Lacustrine, or terrestrial food resources: potential impact on early African Homo Sapiens*. CBP, 2002
-       -   et al: : *Evidence for the Unique Function of DHA during the evolution of the modern Hominid Brain*. Unpublished
Cunningham, E.P., Syrstad, O. *Crossbreeding bos taurus and bos indicus for milk production in the tropics*. FAO of the UN, 1987

Cunnane, Stephen C; Harbige & Crawford. : *The importance of energy and nutrient supply in human brain development.* Nutrition and Health v 9, 1993
Current World Archaeology : *Mons Claudianus & Mons Porphyrites.* (Sources : David Peacock & Valerie Maxfield) No. 8. Nov/Dec 2004
Curtis, Garniss; Swisher, Carl & Lewin, Roger : *Java Man.* Little, Brown & Co. 2001
Cruz-Uribe, Kathryn & Klein, Richard G.: *Faunal Remains from some Middle and Later Stone Age archaeological Sites in South West Africa.* Journal, SWA Scientific Society, 1983
Dalton, Rex : *Ancient footprints found in Mexico.* www.nature.com, 4 July 2005.
Darwin, Charles. *The Origin of Species.* Wordsworth Classics 1998
Davenport, T.R.H.: *South Africa, a Modern History.* Macmillan (S.A.), 1977
Davidson, Basil: *Africa, History of a Continent.* Spring Books, rev. 1972
- - : *The Story of Africa.* Mitchell Beazley, 1984
- - : *The African Past - Chronicles from Antiquity to Modern Times,* Penguin 1966
Dawkins, Richard : *The Blind Watchmaker.* Longman, 1986
- - : *The Extended Phenotype.* Oxford 1999
de Gramont, Sanche: *The Strong Brown God.* Hart-Davis, 1975
Deacon, H.J. : *Guide to Klasies River 2001.* HJ Deacon, Stellenbosch 2001
Deacon, H.J & Deacon, Janette : *Human Beginnings in South Africa.* David Philip 1999
Delegorge, Adulphe : *Travels in Southern Africa,* University of Natal, 1990 (orig. 1847)
Diamond, Jared: *The Rise and Fall of the Third Chimpanzee.* Radius, 1991
- - : *The Shape of Africa.* National Geographic, September 2005
Dick-Read, Robert : *The Phantom Voyagers.* Thurlton Publishing, 2005
Dugard, Martin : *Into Africa.* Bantam Press. 2003.
Duminy, Andrew & Guest, Bill: *Natal and Zululand, from Earliest Times to 1910.* University of Natal - Shuter & Shooter, 1989
Durrani, Nadia : *Flores: Human evolution rewritten.* Current World Archaeology. No 8. 2004
- - : *Çatalhöyük. The most ancient town in the world?* Current World Archaeology. No 8. 2004
Edwards, Stephen John: *Zambezi Odyssey.* Bulpin, 1974
Etler, Dennis: *Homo Erectus in East Asia : Human ancestor or Evolutionary Dead-end?* Athena Review, 2004
Evensberget, Snorre, *Thor Heyerdahl, the Explorer.* Stenersens Forlag, 1994
Fairbridge, Kingsley : *The Story of Kingsley Fairbridge.* Oxford University Press, 1927
Farsi, Sheikh Abdalla Saleh: *Seyyid Said bin Sultan 1804-1856.* Lancers Books, 1986
Felgate, W.S.: *The Tembe Thonga of Natal and Mozambique.* African Studies, University of Natal, 1982
Ferguson, Niall : *Empire - How Britain Made the Modern World.* Allan Lane, 2003
Finkel, Irving: *Games People Keep Playing.* The Illustrated London News, v278-7097
Fock, G.J. & D.: *Felsbilder in Südafrika, Kinderdam und Kalahari.* Böhlau Verlag, 1984
Franzen, Harald. : *Scientists track the origins of Malaria.* Scientific American, 25-6-2001
Gall, Sandy. *The Bushmen of Southern Africa, Slaughter of the Innocent.* Pimlico, 2002
Garlake, Peter S.: *Great Zimbabwe.* Thames & Hudson, 1973
Gayre, R. Gayre of : *The Origin of the Zimbabwean Civilization.* Galaxie Press, 1972
Gerster, Georg: *Tsetse-Fly of the Deadly Sleep.* National Geographic, December 1986
Gibb, H.A.R.(ed.) : *The Travels of Ibn Batuta* vol II. Hakluyt Society, 1959
Gibbon (D.M. Low - ed.) : *The Decline & Fall of the Roman Empire.* Chatto & Windus, 1981
Gibert, J.; Gibert, Ll.; Iglesias.A. & Muestro,E. *Two 'Olduwan' assemblages in the Plio-Pleistocene deposits of the Orce region, southeast Spain.* Antiquity 72, 1998.
Gilbert, Adrian G. & Cotterell, Maurice M.: *The Mayan Prophecies.* Element,1995
Gleick, James: *Chaos, Making a New Science.* Heinemann, 1988
Goodall, Jane van Lawick: *In the Shadow of Man.* Collins, 1971
Goodfellow, Clement Francis : *Great Britain and South African Confederation 1870-1881,* Oxford University Press, Cape Town, 1966
Goodman, Gurasira : *Apollo 11 and the Oldest Rock Art in Africa .* National Museums of Namibia, 1998
Gordon, Ruth & Talbot, Clive J.: *From Dias to Vorster.* Nasou, 1977
Gore, Rick: *Extinctions.* National Geographic, June 1989
- : *The First Steps.* National Geographic, February 1997
Gray, John: *Straw Dogs.* Granta Books, 2002
Gray, Leslie C. & Moseley, William G. : *A geographical perspective on poverty-environment inetractions.* The Geographical Journal, March 2005
Grosset-Grange, H.: *La Côte Africaine dans les Routiers Nautiques Arabes au Moment des Grandes Découvertes.* Azania XIII, 1978.

Grün, R & Beaumont, P. *Border Cave Revisited, a revised ESR chronology.* Jounral of Human
    Evolution , 2001
Gupta, Sunil : *Piracy and trade on the western coast of India (AD1-250).* Azania XLII, 2007
Hakluyt, Richard; ed. Jack Beeching: *Voyages and Discoveries.* Penguin, 1972
Hall, Richard: *Empires of the Monsoon.* Harper Collins, 1996
Hall-Martin, Anthony: Walker, Clive & Bothma, J.du P.: *Kaokoveld, the Last Wilderness.*
    Southern Books, 1988
Hancock, Graham: *Fingerprints of the Gods.* Heinemann, 1995
Hayak, F.A.: *The Road to Serfdom.* Routledge, 1944
Heikell, Rod : *Indian Ocean Cruising Guide.* Imray Laurie Norie & Wilson, 1999
Heinz, Hans-Joachim & Lee, Marshall: *Namkwa, Life among the Bushmen.* Cape, 1978
Henshilwood, CS; Sealy,JC; Yates,R; Cruz-Uribe,K; Goldberg,P; Grine,FE; Klein,RG;
    Poggenpoel,C; van Niekerk,K; Watts,I; *Blombos Cave, Southern Cape, South Africa;*
    Preliminary Report on the 1992-1999 Excavations of the Middle Stone Age Levels.
    Journal of Archaeological Science , 2001.
Henshilwood, Christopher S., et al. : *Emergence of Modern Human Behavior: Middle Stone*
    Age Engravings from South Africa. Science, February 2002.
Herodotus : (Trans. G.C.MaCauley) : *An Account of Egypt.* Gutenberg Etexts , 2000
Heyerdahl, Thor: *The Maldive Mystery.* Allen & Unwin, 1986
  -   -   : *The Tigris Expedition.* Allen & Unwin, 1980
  -   -   : *Fatu-Hiva.* Allen & Unwin, 1974
Highfield, Roger : *Links between humans' ancestors redrawn.* Daily Telegraph, London. 6 May 2008.
Himmelsbach, Dr. Thomas. *The Paleohydrogeology of the Okavango Basin*
    and Makgadikgadi Pan (Botswana) in the Light of Climate Change
    and Regional Tectonics. Bundesanstalt für Geowissenschaften
    und Rohstoffe, 2008
Horton, Mark: : *The Periplus and East Africa.* Azania XXV, 1990.
  -   -   : *The Swahili Corridor .* Scientific American, September 1987
Kramer, Samuel Noah: *Cradle of Civilisation.* Time-Life Books, 1969
Hoste, Skipper. (ed. N.S. Davies) : *Gold Fever.* Pioneer Head, 1977
Huffman, Thomas N.   : *Archaeological Evidence and Conventional Explanations of*
    Southern Bantu Settlement Patterns. Africa, 56 (3), 1986
  -   : *Broederstroom and the Origins of Cattle-keeping in South*
    Africa. African Studies, University of Witwatersrand, 1991
  -   : *Ceramics, Settlements and Late Iron Age Migrations.*
    African Archaeological Review 7, 1989
  -   : *Symbols in Stone.* University of Witwatersrand Press, 1987
  -   : *Snakes and Birds: Expressive Space at Great Zimbabwe.*
    African Studies, University of the Witwatersrand, 1981
  -   : *Handbook to the Iron Age.* University of KwaZulu-Natal Press, 2007
  -   : *Mapungubwe.* Wits University Press, 2005
Humphrey, Nicholas: *The Inner Eye.* Faber & Faber, 1986
Huntingford, G.W.B. ed.: *The Periplus of the Erythraean Sea.* Hakluyt Society, 1980
Howells, William.: *Mankind in the Making.* Doubleday, 1967
Huxley, Elspeth: *Livingstone and his African Journeys.* Weidenfeld & Nicolson, 1974
Hydrographer of the Navy   : *Africa Pilot, vol I.* British Hydrographic Department, 1967
  -   -   : *Africa Pilot, vol II.* British Hydrographic Department, 1977
  -   -   : *Africa Pilot, vol III.* British Hydrographic Department, 1954
  -   -   : *Indonesia Pilot, vol II.* British Hydrographic Department, 1983
  -   -   ; *Mediterranean Pilot, Vol IV.* British Hydrographic Department, 1955
  -   -   : *Red Sea and Gulf of Aden Pilot.* British Hydrographic Department, 1980
  -   -   : *South America Pilot, Vol I.* British Hydrographic Department, 1975
  -   -   : *South Indian Ocean Pilot.* British Hydrographic Department, 1971
  -   -   : *West Coast of India Pilot.* British Hydrographic Department, 1975
  -   -   : *Ocean Passages for the World.* British Hydrographic Department, 1973
Ingman, Max : *Mitochondrial DNA Clarifies Human Evolution.* Bioscience Productions. 2001
Inman, Mason : *Human brains enjoy ongoing evolution.* New Scientist news service, 15 Sep 2005
Inskeep, Ray : *The Problem of Bantu Origins.* Duckworth 1973
Isaac, Glynn: *Visitors' Guide to the Olorgesaile Prehistoric Site.* Museums of Kenya, 1985
Isaacs, Nathaniel. (ed. Herrman, Louis) *Travels and Adventures in Eastern Africa.*
    Van Riebeeck Society, 1937
Jabavu, Noni : *The Ochre People.* John Murray, 1963

Jacobson, L.: *The Archaeology of the Kavango*. Journal, SWA Scientific Society, 1987
- : *The Brandberg*. Rössing Magazine, Dec., 1981
Johanson, Donald C. & Shreeve, James : *Lucy's Child*. Viking, 1990
Johanson, Donald C.: *The Dawn of Humans, Face to Face with Lucy's Family*. National Geographic, March 1996.
Jakubowski, Peter. : *The Cosmic Carousel of Life* Naturics Foundation. 2003
Kaner, Simon : *The Oldest Pottery in the World*. Current World Archaeology. v 1. 2003.
Katanekwa, N.M.: *Some Early Iron Age sites from the Machili Valley of South Western Zambia*   Azania XIII, 1978
Keller, Werner: *The Bible as History*. Hodder & Stoughton, 1956
Kinehan, J.: *The Stratigraphy and Lithic Assemblages of Falls Rock Shelter, Western Damaraland, Namibia*. Cimbebasia (B) 4 (2), 1984
Kingsley, Mary : *Travels in West Africa*. MacMillan. 1897
Kinver, Mark : *Tools unlock secrets of early man*. BBC News website, 14-12-2005
Kirkman, James: *The Early History of Oman in East Africa*. Journal of Oman Studies, 1983
- - : *Gedi*. Museums of Kenya, 1975
- - : *Fort Jesus*. Museums of Kenya, 1981
Kislev, Mordechai; Hartmann, Anat; Bar0Yosef, Ofer : *Early Domesticated Fig in the Jordan Valley*. Science, 2-6-2006
Kislev, M.E.; Nadel, D.; Carmi, L. : *Epipaleolithic (19,000 BP) Cereal and Fruit Diet at Ohalo II, Sea of Galilee, Israel*. Review of Paleobotany & Palymology. V.73 1992
Klein, Richard G., *Whither the Neanderthals?*. Science, v.299, 2003
- - & Edgar, Blake. *The Dawn of Human Culture*. John Wiley & Sons, 2002
Knehtl. Irena.: *The Voyages of Zheng He*. Yemen Times, 8 September 2005
Latham, Ronald [translator]: *The Travels of Marco Polo*. Folio Society, 1968
Larick, Roy; Ciochon, Russell & Zaim, Yahdi: *Homo erectus and the Emergence of Sunda in the Tethis Realm*. Athena Review v 4, No 1. 2004
LaViolette, Paul A.: *Earth under Fire*. Bear & Co, 2005
- - . : *Evidence for a Global Warming at the Termination I Boundary and Its Possible Cosmic Dust Cause* .The Starburst Foundation, Fresno, USA, 2005
Leakey, Mary: *Disclosing the Past*. Weidenfeld & Nicolson, 1984
- - : *Olduvai Gorge, my Search for Early Man*. Collins 1979.
Leakey, Meave: *The Dawn of Humans, The Farthest Horizon*. National Geographic, September 1995
Leakey, Richard E.: *Skull 1470*. National Geographic, June 1973
- - & Lewin, Roger: *Origins*. Macdonald & Jane's, 1977
- - & Walker, Alan: *Homo Erectus Unearthed*. National Geographic, November, 1985
- - & Lewin, Roger: *Origins Reconsidered*. Little, Brown, 1992
Lemonick, Michael & Dorfman, Andrea : *Father of us All?* Time Magazine 22-7-2002
Lewin, Roger: : *Complexity, Life at the Edge of Chaos*. Dent, 1993
Lewis, David. *We, the Navigators*. University of Hawaii Press, 1994
Lewis-Williams, J. D. : *The Economy and Social Context of Southern San Rock Art*. Current Anthropology v23.4., 1982
Lewis-Williams, David.: *Images of Mystery, Rock Art of the Drakensberg*. Double Storey Books, 2003
- - - : *The Mind in the Cave*. Thames & Hudson, 2002
- - - & Pearce, David : *Inside the Neolithic Mind*. Thames & Hudson, 2005
- - - & Dowson : Thomas: *Images of Power*, Southern Books, 1989
Leyland, J.: *Adventures in the Far Interior of South Africa*. Routledge, 1866
Lhote, Henri: *Oasis of Art in the Sahara*. National Geographic, Aug. 1987
Liebenberg, Doyle P.: *The Drakensberg of Natal*. Bulpin, 1972
Liveing, Edward. *Across the Congo*. Witherby, 1962
Livingstone, David : *Missionary Travels and Researches in South Africa*. 1857. (Gutenberg Etexts 1997)
- - : *A Popular Account of Dr. David Livingstone's Expedition to the Zambesi and its Tributaries & the Discovery of Lakes Shirwa & Nyassa*. Johh Murray 1894
- - (Ed. Waller, Horace.) : *The Last Journals of David Livingstone in Central Africa from 1865 to his Death*. (2 vols.) John Murray 1874
Lock, Ron: *Blood on the Painted Mountain*. Greenhill Books, 1995
Londhe, Sushama. *Seafaring in Ancient India*. A tribute to Hinduism - WWW, 2001
Lovelock, James: *The Ages of Gaia*. Oxford, 1988
- - : *The Revenge of Gaia*. Allen Lane, 2006
Lüthi, Dieter, et al : *High-resolution carbos dioxide concentration record 650,000-800,000 years before present*. Nature, 15 May 2008
Mabulla, Audax Z. P. : *The Rock Art of Mara Region, Tanzania*. Azania XL, 2005

Mack, John : *The Land Viewed from the Sea.* Azania XLII. 2007
Maclean, Charles Rawden: *The Natal Papers of John Ross.* Univ. of Natal, 1992
Macnair, James I.: *Livingstone's Travels.* Dent, 1956
Maggs, Tim: *Mzonjani and the Beginning of the Iron Age in Natal.* Natal Museum 24(1), 1981
-   - : *The Iron Age South of the Zambezi.* A.A.Balkema, 1984
Majid al-Najdi, Ahmad ibn (translator and editor : Tibbets, Gerald R.) : *Arab Navigation in the Indian Ocean before the coming of the Portuguese.* Royal Asiatic Society, 1971
Mandela, Nelson. *Long Walk to Freedom.* Macdonald & Purnell, 1995
Manwell, Clyde & Baker: *Chemical Classification of Cattle.* Genet.11, University of Adelaide, 1980
Marais, Eugene: *The Soul of the Ape.* Human & Rousseau, 1969
Marscher, Alan P. et al : *The inner jet of an active galactic nucleus as revealed by a radio-to-γ ray outburst.* Nature, vol 452. 24-4-2008
Marsh, David : *Waters Edge Man: the new paradigm currently displacing the Savannah Theory of Human Origins.* Nutrition & Health . v. 14.2
-   - : *Role of the Essential Fatty Acids in the Evolution of the Modern Human Brain.* Positive Health Magazine, May 2001
-   - : *The Origins of Diversity: Darwin's Conditions and Epigenetic Variations* Nutrition & Health, vol 19, 2007
Marshall, John & Ritchie, Claire: *Where are the JU/WASI of Nyae Nyae?.* African Studies, University of Cape Town 1984
Maslin, Mark; Trauth, Martin; Christensen, Beth : *A Changing Climate for Human Evolution.* Geotimes. Septermber 2005.
Mathews, Robert. *Meteor Clue to end of Middle East Civilisations.* Sunday Telegraph 4-11-2001
Matthews, Samuel W.: *Ice on the World.* National Geographic, Jan. 1987
Matthiessen, Peter: *The Tree where Man was Born.* Collins, 1972
Maylam, Paul: *A History of the African People of South Africa.* David Philip, 1986
McCarrison Society . www.mccarrisonsociety.org.uk
McCrindle, J.W.: *The Christian Topography of Cosmas, an Egyptian Monk.* Hakluyt Society, 1897
McLaren, Angus: *Reproductive Rituals.* Methuen, 1984
McPhun, Delwyn : *East Africa Pilot.* Imray Laurie Norie & Wilson, 1998
Meier, David L, : *Exhaust Inspection,*. Nature, 24-4-2008
Meinel, Aden B.; Meinel. Marjorie P.; Meinel, B.; Drach-Meinel, D. : *Fingerprints in Ice: A Cosmic Encounter, a Cat's Eye, and Origin of Modern Humans.* Theoretical Archeological Group Conference, Sheffield University, December 2005.
Mendelssohn, Kurt: *Riddle of the Pyramids.* Praeger, 1974
Merfield, Fred G. & Miller, Harry: *Gorillas were my Neighbours.* Longmans, 1956
Merrick, H.V.: *Visitors Guide to the Hyrax Hill Site.* Museums of Kenya, 1983
Milliken, Sarah: *Out of Africa or Out of Asia? The colonization of Europe by Homo erectus.* Athena Review. V4, No 1. 2004
Mitford, Bertram: *Through the Zulu Country.* Kegan Paul, 1883
Monbiot, George : *No Man's Land.* Macmillan, 1994
Montgomery, Brian : *Shenton of Singapore.* Leo Cooper, 1984
Montgomery, Denis : *The Reflected Face of Africa.* African Insight, 1988
-   - : *Two Shores of the Ocean.* Malvern Publishing, 1992
Moorehead, Alan: *The White Nile.* Hamish Hamilton, 1960
-   -   - : *The Blue Nile.* Hamish Hamilton, 1962
Morais, João: *Mozambican Archaeology: Past and Present.* African Archaeological Review, 2, 1984
Morgan, Elaine: *The Descent of Woman.* Souvenir Press, 1972
-   - : *The Aquatic Ape.* Souvenir Press, 1982
-   - : *The Scars of Evolution.* Souvenir Press, 1990
-   - : *The Aquatic Ape Hypothesis,* Souvenir Press.1997
Morgan, Ekaine & Davies, Stephen. *Red Sea Pilot.* Imray, Laurie, Norie & Wilson, 2002.
Morris, Desmond: *The Naked Ape.* Cape, 1967
-   - : *The Human Zoo.* Cape, 1969
-   - : *Manwatching.* Cape, 1977
-   - : *The Naked Woman.* Cape, 2004
Morris, Donald R.: *The Washing of the Spears.* Cape, 1966
Morwood, M.J., et al : *Archaeology and age of new hominin from Flores in eastern Indonesia.* Nature, 27-11-2004
Murray, H.J.R.: *History of Board Games other than Chess.* 1952
Mutwa, Credo: *Indaba, my Children.* Blue Crane, 1964
NASA : *NASA Achieves Breakthrough In Black Hole Simulation* - News release 18/4/2006

National Geographic Society: *Where did Columbus Discover America?*. Nov. 1986
- - - : *Does Living a Stone Age Life Cut Cancer Risk?*. August, 1993
Nature : Sundry original papers, including Dart, Johanson, Leakey etc : www.nature.com/nature/ancestor
Newson-Smith, Sue: *Quest, the Story of Stanley and Livingstone*. Arlington Books, 1978
Nicholson, Ward: *Longevity and Health in Ancient Palaeolithic v Neolithic Peoples*. www.beyondveg.com , 1999
Nimmo, W.P, Hay & Mitchell (pub.): *The Life and Travels of Mungo Park*. Edinburgh, n/d
O'Brien, Frederick : *Mystic Isles of the South Seas*. Hodder & Stoughton, 1921
Oberholster, J.J.: *The Historical Monuments of South Africa*. Nat. Monuments Council, 1972
O'Connor, Anthony: *Poverty in Africa*. Belhaven Press, 1991
Oliver, Roland: *The African Experience*. Weidenfeld & Nicolson, 1991
- - & Fage, J.D.: *A Short History of Africa*. Penguin, 1962
Oppenheimer, Stephen : *Out of Eden, the Peopling of the World*. Constable. 2003
Owens, Mark & Delia: *Cry of the Kalahari*. Collins, 1985
Paice, Edward : *Lost Lion of Empire*. Harper Collins 2001
Pakenham, Thomas: *The Boer War*. Weidenfeld & Nicolson, 1979
- - : *The Scramble for Africa*. Weidenfeld & Nicholson, 1991
- - : *The Mountains of Rasselas*. Weidenfeld & Niicolson. 1998
Parfitt, Simon. *Pakefield: a weekend to remember*. British Archaeology. V 86. Jan-Feb 2006
Parker, Eduard: *High Road or Road to Ruin?* Optima, v 38/2, 1992
Parkington, John : *The Mantis, the Eland & the Hunter*. Krakadouw Trust, 2002
- : *Cederberg Rock Paintings*. Krakadouw Trust, 2003
- : *Shorelines, Strandlopers and Shell Middens*. Krakadouw Trust 2006
Parsons, Janet Wagner. *The Livingstones at Kolobeng 1847-1852*. Botswana Society & Pula Press, 1997
Parsons, Neil : *Botswana History Pages*. WWW 2000
Pearse, R.O.: *Barrier of Spears*. Howard Timmins, 1973
Pennisi, Elizabeth : *Genomes Throw Kinks in Timing of Chimp-Human Split*. Science, v 312. 19 May 2006
Phillipson, David W. *African Archaeology, 2nd Edition*. Cambridge, 1993
Phillipson, Laurel. *Ancient Gold Working at Aksum*. Azania XLI, 2006
Potts, Richard et al : *Small Mid-Pleistocene Hominin Associated with East Africa Acheulean Technology*. Science, 2 July 2004
Putnam, John J.: *The Search for Modern Humans*. National Geographic, Oct. 1988
Ransford, Oliver: *David Livingstone, the Dark Interior*. John Murray, 1978
Rasmussen, R.Kent: *Migrant Kingdom*. Rex Collings, 1978
Raven-Hart, R.: *Before van Riebeeck*. Struik, 1967
Ravenstein, E.G.: *The Voyages of Diogo Cão and Bartholomeu Dias 1482-1488*. The Royal Geographical Society Journal, 1900.
Reader, John : *Africa, a Biography of a Continent*. Penguin 1997
Redfield, T.F.: *A kinematic model for Afar Depression lithospheric thinning and its implications for hominid evolution : an exercise in plate tectonic paleoanthropology*. Geological Society of America, 2002
Redfield, T.F.; Wheeler. W.H.; Often, M. : *A kinematic model for the development of the Afar Depression and its paleogeographical implications*. Pre-publication paper. 2003
Reik, W., et al. : *Epigenic reprogramming in mammalian development*. Science, 2001.
Rigaud, Jean-Philippe: *Art Treasures from the Ice-Age, Lascaux Cave*. National Geographic, October, 1988
Rincon, Paul : *Footprints of 'First Americans'* . BBC News, 5 July 2005
- : *Hobbit was 'not a diseased human'* . BBC News 3 March 2005
Robertshaw, Peter : *Early Pastoralists of South-western Kenya*. BIEA, 1990
Roebroeks, Wil. *Archaeology: Life on the Costa del Cromer*. Nature, 15-12-2005
Roede, Machteld; Wind, Jan; Patrick, John & Reynolds, Vernon [eds.]: *The Aquatic Ape: Fact or Fiction*. Souvenir Press, 1991
Roodt, Dan. *The Scourge of the ANC*. Uitgewers Praag. 2004
Rosenblum, Mort & Williamson, Doug: *Squandering Eden*. Bodley Head, 1988
Russell, Peter. *Prince Henry 'the Navigator', A Life*. Yale University 2000
Sadr, Karim. *The First Herders at the Cape of Good Hope*. African Archaeological Review 15/2, 1998
Sample, Ian : *Neanderthal DNA reveals human divergence*. Guardian Unlimited. 16-11-2006
Sandelowsky, B.H.: *Archaeology in Namibia*. American Scientist v.71, 1983
Sanders, Peter: *Moshoeshoe, Chief of the Sotho*. Heinemann-Philip, 1975
Saugstad, Letten F. : *Are neurodegenerative disorder and psychotic manifestations avoidable brain dysfunctions with adequate dietary omega-3?* Nutrition & Health 18-2 (2006)
Schapera, I.: *Married Life in an African Tribe*. Faber, 1940
Schapera, I (ed.) *David Livingstone South African Papers 1849-1853*. Van Riebeeck Society, 1974
Schapera, I (ed.) *Bantu-Speaking Tribes of South Africa*. 1937

Schilling, Govert. : *Do Gamma Ray Bursts Always Line Up with Galaxies?* Science v 313 - 5788, 2006
Schoeman, P.J. (Prof.) : *Hunters of the Desert Land.* Howard Timmins, 1957
Schoff, William, H. : *The Periplus of the Erythraean Sea.* Longmans Green, NY. 1912
Schwandler, Jakob, et al. : *A tentative chronology for the EPICA Dome Concordia ice core.* Geophysical Rearch Letters. V.28 no.22, 2001
Segal, Ronald : *Islam's Black Slaves.* Atlantic Books, 2001
Selous, Frederick Courtney: *Travel and Adventure in South-East Africa.* Rowland Ward, 1893
Semino, Ornello ; Santachiara-Benerecetti, A.Silvano ; Falaschi, Francisco ; Cavalli-Sforza, L.Luca ; Underhill, Peter A. : *Ethiopians and Khoisan share the deepest Clades of the Human Y-chromosome Phylogeny.* American Journal of Human Genetics, 2002.
Shlovskii, I.S. & Sagan, Carl: *Intelligent Life in the Universe.* Holden-Day, 1966.
Sinclair, Paul : *Chibuene - An Early Trading Site in Southern Mozambique.* Paideuma 28, 1982
- - : *Ethno-Archaeological Surveys of the Save River Valley, South Central Mozambique.* African Studies Programme, Uppsala. 1985
Smillie, Shaun : *Survival was simply about luck.* The Mercury, Durban, 25 April 2008.
Soderberg, A.M., et al. : *An extremely luminous X-ray outburst at the birth of a supernova.* Nature, 22-5-2008
Soper, Robert. *Nyanga, Ancient fields, settlements and agricultural history in Zimbabwe.* BIEA., 2002
Sparrman, Anders : *A Voyage to the Cape etc. and to the Country of the Hottentots and the Caffres from the year 1772-1776.* Robinson, London. 1785
Speke, John Hanning. *Journal of the Discovery of the Source of the Nile.* Blackwood, 1863
Spindler, Konrad: *The Man in the Ice.* Weidenfeld & Nicolson, 1994
Stanley, Henry Morton : *Through the Dark Continent.* Dover Publications. 1988 (orig1899)
Steenkamp, Willem : *De Klerk stands up to 'unfair discrimination'.* Sunday Argus, 6-11-2005.
Steyn, H.P. & du Pisani, E.: *Grass Seeds, Game and Goats: an Overview of Dama Subsistence.* Journal, SWA Scientific Society, 1985
Stone, Richard : *Java Man's First Tools.* Science magazine, 21 April 2006
Stringer, Chris : *Modern Human Origins.* The Royal Society, April 2002
- - : *The First Humans North of the Alps.* British Archaeology v 86, Jan-Feb 2006
Stringer, Chris & McKie, Robin. *African Exodus, the Origins of Modern Humanity.* Jonathan Cape, 1996
Stringer, Chris & Gamble, Clive. *In Search of the Neanderthals.* Thames & Hudson, 1993
Stuart, James & Malcolm, D.McK. (Eds): *The Diary of Henry Francis Fynn.* Shuter and Shooter, 1969
Subrahmanyam, Sangay : *The Career and Legend of Vasco da Gama.* Cambridge University Press, 1997
Summers, Roger: *Ancient Ruins and Vanished Civilisations of Southern Africa.* Bulpin, 1971
Sutton, J.E.G: *A Thousand Years in East Africa.* British Institute in Eastern Africa, 1990
- : *Engaruka and its Waters.* Azania XIII, 1978
- : *Towards a History of Cultivating the Fields.* Azania XXIV, 1989
- [Ed]: *The Growth of Farming Communities in Africa from the Equator Southwards.* Azania [BIEA], 1996
Taylor, Stephen : *Livingstone's Tribe.* Harper Collins, 1999
Templeton, Alan : *Recent finds in archaeology.* Nature. March 2002
Theal, George McCall: : *Records of South-Eastern Africa,.* Government of the Cape Colony, 1899
- - - : *History of South Africa* , Vols I-XI (facsimile), Struik, 1964
Theroux, Paul: *The Happy Isles of Oceania.* Hamish Hamilton, 1992
Thomas, Elizabeth Marshall: *The Harmless People.* Secker & Warburg, 1959
Thompson, George : *Travels and Adventures in Southern Africa.* Henry Colburn, 1827
Thubron, Colin. : *Shadow of the Silk Road.* Chatto & Windus, 2006
Tishkoff, Sarah H., et al. : *History of Click-Speaking Populations of Africa Inferred from mtDNA and Y Chromosome Genetic Variation.* Molecular Biology and Evolution 24-10, 2007
Tobias, Phillip V .: *The Peoples of Africa South of the Sahara.* Clarendon Press, 1966
- : *On the Increasing Stature of the Bushmen.* Anthropos v 57, 1962
- : *Recent Human Biological Studies in Southern Africa with Special Reference to Negroes and Khoisans.* Royal Society of South Africa. 40-3, 1972
- : *Into the Past.* Picador Africa, 2005.
Trauth, Martin; Maslin, Mark; Delno, Alan; Streckner, Manfred : *Late Cenezoic Moisture History of East Africa.* Science v.309. September 2005.
Uhlig, Robert: *Mega-Neutron blast 'killed off dinosaurs'.* The Daily Telegraph, December 1996
UNESCO - Mokhtar, G., Hrbek, I. (Eds): *General History of Africa* (Abridged Edn). Univ. of California Press, James Currey & UNESCO, 1990
Uyanker,B., Reich, W., Yar, A., Kothes, R., Fürst, E. : *Is the Cygnus Loop two supernova remnants?* Astronomy and Astrophysics, 2002.
Vallee, Jacques : *Dimensions, A casebook of Alien Contact.* Souvenir Press. 1988
van Grunderbeek, Marie-Claude. *Chronologie de l'Age du Fer Ancien au Burundi, au Rwanda et dans le Région des Grands Lacs*, AZANIA XXVII, 1992

van Rijn, Hans & Hutt, Graham : *North Africa*. Imray Laurie Norie & Wilson, 2000
van Schalkwyk, Len: *Settlement Shifts and Socio-economic Transformations in Early Agriculturalist Economies in the Lower Thukela Basin*. AZANIA XXIX-XXX, 1996
- - & Greenfield, Haskel J. *Intra-settlement Social and Economic Organisation of Early Iron Age Farming Communities in Southern Africa: a view from Ndondwane*. Azani XXXVIII, 2003
Vansina, Jan: *Western Bantu Expansion*. Journal of African History v 25, 1984
- - : *A Slow Revolution : Farming in sub-Tropical Africa*. AZANIA XXIX-XXX ,1996
Verhagen, Marc & Puesch, Peirre-Francois : *Hominid Lifestyle and Diet Reconsidered: Paleo-Environmental and Comparative Data*. Human Evolution 15, 2000
Vernet, Thomas. *Le commerce des eclaves sur le côte swahili*, 1500 - 1750. Azania XXXVIII, 2003
Villiers, Alan. : *Sons of Sinbad*. Hodder & Stoughton, 1940
- - : *The Set of the Sails*, Hodder & Stoughton, 1949
Wainwright, G.A. : *Cosmas and the Gold Trade of Fazogli*. Man, vol 42, no. 30. June 1942
Walker, Iain : *Hadramis, Shimalis and Muwalladim: Negotiating Cosmopolitan Identities between the Swahili Coast and Southern Yemen*. Journal of Eastern African Studies, March 2008
Wandibba, Simiyu: *Ancient and Modern Ceramic Traditions in the Lake Victoria Basin of Kenya*. Azania XXV 1990
Watson, Lyall : *Earthworks*. Hodder & Stoughton, 1986
- - : *Dark Nature*, Hodder & Stoughton, 1995
- - : *Lightning Bird*, Hodder & Stoughton, 1982
- - : *Elephantoms*, Norton & Co., 2002
Weinberg, Paul : *Once We were Hunters*. David Philip, 2000
Wendt, C.E.: *'Art Mobilier' from the Apollo 11 Cave, South West Africa*. South African Archaeological Bulletin 31, 1976
Werz, B.E.J.S. & Flemming, N.C. *Discovery in Table Bay of the oldest handaxes yet found demonstrates preservation of hominid artefacts on the continental shelf*. South African Journal of Science, May/June 2001.
Wheeler, Sir Mortimer: *The Indus Civilisation*. Cambridge, 1968
West Africa, 16-22 September 1966. *Wanted: skill, speed, strategy*.
White, Randall. *Personal Ornaments from the Grotte du Rennes at Arcy-sur-Cure*. Athena Review 2001
White, Tim D. et al : *Pleistocene Homo sapiens from Middle Awash, Ethiopia*. Nature, 12 June 2003
Whitelaw, Gavin & Moon, Michael : *The ceramics and distribution of pioneer agriculturalists in KwaZulu-Natal*. Natal Museum Journal of Humanities, 1996
Wilding, Richard: *The Shorefolk*. Fort Jesus Museum, 1987
Willcox, A.R.: *The Drakensberg Bushmen and their Art*. Drakensberg Publications, 1984
Wilfiord, John Noble : *Mutation Cited in Evolution*. New York Times, 24 March 2004
Wills, A.J.: *An Introduction to the History of Central Africa*. Oxford, 1973
Wilson, Ian: *The Exodus Enigma*. Weidenfeld & Nicolson, 1985
Wilson, Monica & Thompson, Leonard (eds.): *Oxford History of South Africa*. Oxford, 1969
Withnell, Allan. *The Nature and Importance of Our Prehistoric Diet*. Nurition & Health v.17 No.4, 2004
Wong, Kate : *The Littlest Human*. Scientific American, Feb. 2005
Wood, Bernard : *Hominid Revelations from Chad* : Nature 418, 2002
Wrigley, C.C.: *Bananas in Buganda*. Azania XXIV, 1989
Young, Anthony. *Poverty, hunger and population policy: linking Cairo with Johannesburg*. The Geographical Journal, March 2005
Young, E.D.: *Nyassa, a Journal of Adventures*. John Murray, 1877.
Züchner, Christian : *Grotte Chauvet Archaeololically Dated*. IRAC Conference papers. 1998

GENERAL INDEX

Aborigines 79, 108, 110
Abydos 131, 132
Acheulian 29, 49
Adam 48, 280
Afar 285
African Eve 1, 4, 14, 15, 18, 19, 35, 84, 279
Afrikaners 73, 107, 274
Afro-Asiatic 51, 57, 82, 167
Aguada da Boa Paz 263, 269
Alexander the Great 151
Algarve 250, 254
Anatolia 23, 24, 26, 27, 38, 133, 134
Andrew Anderson 244
Andrew Collins 16, 146
Angra Pequena 257
Antarctic 279
antelopes 28, 34, 73, 74, 117, 181
Anthropic Cosmological Principle 279
Aqaba 135
Ardrey 45, 279
Assyrians 133, 135, 143, 151
Athena Review 25, 279-281, 283, 284, 287
Australia 79, 108-110, 128
Australopithecine 15
Australopithecus 56, 130, 227
Axelson 8, 104, 184, 188, 250, 253, 260, 261, 263, 273, 279
Axum 125, 139, 154, 206, 254
Azania 7, 58, 59, 65, 68, 155, 157, 158, 165, 167, 188, 224, 279-287
baboons 23, 46, 73, 112, 136
Babylonians 133
Bagamoyo 159, 178, 182, 183
bananas 7, 53, 63-65, 70, 183, 287
Bantu 5, 19, 57, 66, 68, 75, 85-92, 94, 96, 100-102, 105-107, 109-111, 148, 150, 156, 158, 159, 162, 167, 169,
177, 191-194, 198, 199, 201, 203, 208, 209, 212, 215-218, 227, 229, 230, 232, 235, 238,
243, 254, 269, 282, 285, 287
Bartolomeo Dias 103, 256, 257, 260, 264
Basil Davidson 106, 166, 236
Bazaruto Bay 188, 222-224
Benin 54, 254, 256
Berber 82, 83, 147, 254
Berenice 134
Bering Strait 128
Bible 54, 136, 137, 283
bipedal 15
Blackburn pottery 229
Blombos Cave 191, 282
Blue Nile 139, 227, 284
Bogoria 52, 58
Bojador 252, 253
Border Cave 191, 279, 282
bos indicus 55, 119-122, 125, 280
bos taurus 55, 68, 118-122, 125, 280
Botswana 42, 72, 73, 81, 86, 92, 93, 99, 231-233, 236, 242, 244, 282, 285
Brandberg 283
Brazil 1, 187, 261, 270
Broederstroom 282

bronze. 43, 132, 135, 141, 154, 232, 235
Brunet. 280
camels. 39, 122, 134, 135, 150
Cameroon. 145, 198
canoes. 52, 53, 64, 128-130, 134, 146, 149, 183, 187
Canton. 256
Cão. 256, 257, 285
Cape Correntes. 164, 174, 176, 182, 187, 225, 252, 269
Cape Cross. 257
Cape of Good Hope. 78, 81, 93, 103, 150, 174, 177, 194, 204, 247, 257, 261, 264, 265, 271, 274, 285
Cape Town. 14, 42, 103, 105, 106, 108, 274, 281, 284
Capricorn. 145, 174, 198, 225, 226
caravel. 253
carbon-dating. 230
Caribbean. 146, 150, 187, 247, 248, 270, 272
cassava. 63
Çatal Hüyük. 23, 26
cattle. 5, 20, 27, 28, 34, 38, 42, 43, 47, 53, 55, 57, 59, 62-66, 68, 71, 72, 92, 94, 96, 99, 101, 102, 104-107, 109, 117-126, 138, 149, 156, 168, 169, 195-197, 201-203, 207-209, 215-219, 226-228, 231, 232, 238, 240, 241, 244, 254, 258, 269, 282, 284
cattle-cult. 101, 124, 195, 217, 218
Central America. 129, 132, 146, 256
Central Cattle Pattern. 216-219
Cetshwayo. 121, 243
Ceuta. 250, 251
Chad. 43, 200, 280, 287
Chami. 14, 57, 66-69, 82, 83, 92, 147, 156, 158, 161, 165-171, 174, 200, 228, 280
Chauvet. 287
Cherangani Hills. 55
Chibuene. 159, 188, 197, 222-225, 236, 286
chimpanzees. 46
China. 35, 140, 163, 164, 184, 186, 256, 262, 268, 274
Chinese. 132, 162-164, 184, 268, 269
Christian. 105, 149, 150, 156, 161, 162, 164, 169, 170, 173, 174, 185, 190, 193, 200, 227, 249, 254, 255, 267, 284, 287
citrus. 63
civilisation. 5, 13, 21, 23, 26, 30, 31, 33, 35, 36, 39, 41-48, 52, 60, 61, 63, 68, 75, 80, 82, 112, 122, 130, 131, 133, 134, 142, 152, 161, 166, 169, 172, 210, 230, 242, 244, 247, 248, 282, 287
clients. 28, 61, 80, 96, 108, 109, 125, 148, 156, 194, 205, 211, 218, 238, 257
clientship. 28, 29, 31, 42, 45, 47, 52, 56, 61, 75, 81, 96, 100, 105, 119, 165, 218, 219
climate. . . 21, 26-28, 32, 41, 42, 49, 55, 57, 62, 80, 96, 110, 123, 125, 135, 141, 142, 146, 156, 191, 195, 198, 210, 219, 279, 282, 284
climate change. 26, 27, 41, 282
coconuts. 181, 183
Columbus. 127, 247, 252, 256, 260, 285
Comores. 50, 149, 162, 186
Congo. 20, 43, 54, 57, 76, 90, 92, 120, 123, 124, 172, 198, 212, 216-219, 228, 250, 257, 261, 273, 279, 283
Congo Basin. 216-218, 228
Congo River. 172, 257, 273
copper. 43, 88, 102, 105, 132, 135, 136, 141, 154, 163, 186, 226, 231, 232, 235, 269
cosmic dust. 283
cosmic radiation. 15, 16, 35, 36, 48, 119
cotton. 24, 141, 163, 180-182, 235
cowrie shells. 131, 196
Cro-Magnon. 191
Cushitic. 65, 69, 89-92, 120, 125, 156, 159, 167, 169, 193, 196, 203, 218
Cygnus. 16, 48, 280, 286
Cygnus X-3. 280
da Gama. 8, 19, 103, 104, 120, 189, 222, 259-269, 279, 286
Dart. 285
Darwin. 281
Delagoa Bay. 108, 176, 177, 204, 225
dialect. 72

difaqane... 204
diffusion... 19, 21, 31, 41, 44, 51, 57, 61, 66, 91, 93-98, 100, 103, 125, 150
Dingane... 243
Diu... 271
DNA... 85, 280, 282, 285
Dordogne... 118
Dorobo... 55
Drakensberg... 79, 191, 192, 194, 195, 238, 283, 287
dugout canoes... 146
Durban... 14, 193, 194, 196, 197, 286
Dutch East India Company... 107
Eannes... 253
Early Iron Age... 51, 68, 69, 98, 150, 156, 158, 159, 161, 162, 168, 169, 193-200, 208, 212, 219, 222, 223, 226, 229, 280, 283, 287
Early Stone Age... 29, 49, 54, 130
East Indies... 63, 271, 274
Egypt... 20, 36-38, 43, 54, 55, 60, 66, 118, 120, 125, 131-135, 137, 140, 141, 143, 144, 147, 151, 152, 154, 158, 164, 168, 174, 247, 250, 254, 258, 270, 271, 282
Elementeita... 49
elephants... 113, 154, 186, 280
Elgeyo Escarpment... 59
Elizabeth Marshall Thomas... 74, 81
Elmina... 253, 256, 277
Enkwalini... 192
epigenetic... 284
Ethiopia... 42-44, 52, 55-57, 59-61, 65, 89-91, 120, 125, 137, 151, 154, 167-169, 174, 184, 190, 206, 210, 226, 227, 240, 254, 255, 258, 275, 279, 286, 287
Euphrates... 131, 134, 142
Eurasia... 15, 118, 249
Eve... 1, 4, 14, 15, 18, 19, 35, 84, 279
extinctions... 16, 281
Ezion-Geber... 137
Falls Rock Shelter... 97, 98, 283
Felix Chami... 14, 57, 66, 67, 69, 82, 92, 147, 158, 165, 167, 200, 228
fire... 46, 95, 104, 106, 111, 262, 283
fishing... 32, 41, 42, 53, 62, 64, 78, 87, 101, 123, 129, 130, 134, 156, 187, 199, 216, 252, 257, 267
fishtraps... 155
Flores... 130, 281, 284
Gaia... 283
Galla... 203, 206
Gambia... 253
gaming boards... 50
Ganges... 153
Garamantian... 39, 42, 43
Gavin Whitelaw... 14, 193, 196, 197, 199, 212
Gedi... 203, 283
genes... 16, 17, 80, 84, 109, 122, 129, 147, 195
genetic imperative... 36
Genoa... 250, 251
Ghana... 14, 146, 248, 253, 277
GiKwe... 76, 77
global warming... 283
gold... 7, 8, 132, 135-139, 154, 160, 162-164, 171, 178, 180, 182, 184, 186, 211, 221, 223-225, 227, 228, 231-237, 240, 241, 249-251, 253-255, 263, 270, 271, 282, 285, 287
gorillas... 46, 113, 284
Great Rift Valley... 41, 49, 52, 54, 56, 59, 60, 65, 124, 125, 135, 152, 170, 196, 201, 210, 227, 228
Great Zimbabwe... 21, 62, 164, 211, 223, 224, 229, 230, 232, 233, 235-245, 281, 282
Greeks... 13, 132, 140, 147, 151, 167
Hadza... 55
Hanno... 145, 147
Harappa... 134
Herero... 87, 107
Herodotus... 143, 145, 152, 167, 168, 282

Heyerdahl. 130, 281, 282
highveld. 79, 195, 201, 204, 216, 217, 228, 243
Himba. 111
Hindu. 224, 270
Hittites. 133, 135, 151
Hobbit. 285
Holocene wet phase. 41
homesteads. 49, 53, 62, 216, 217
Homo. 15, 16, 30, 33, 35, 36, 41, 49, 69, 85, 130, 279-281, 283, 284, 287
Homo erectus. 15, 16, 30, 35, 49, 69, 130, 280, 281, 283, 284
Homo sapiens. 16, 35, 41, 85, 279, 280, 287
Hormos. 134
Horn of Africa. 7, 39, 44, 55, 60, 65, 119, 125, 128, 130, 135, 139-141, 154, 167, 174, 196
horses. 28, 39, 42, 43, 118, 122, 124, 125
Horton. 159, 174, 282
Hottentot. 71, 87, 107
Huffman. 14, 21, 40, 201, 208, 210, 214-217, 219, 229, 233, 235-237, 239, 242, 243, 282
hunter-gatherers. 20, 23, 26, 29, 31, 45, 48, 55, 57, 61, 79, 93-96, 100, 129, 192
hunter-gathering. 27, 33, 54, 71, 82, 86, 100, 107
hunting. 26-28, 33, 46, 48, 61, 62, 81, 86, 89, 91, 94, 101, 129, 142, 154, 171, 279
huts. 50, 53, 69, 101, 110, 111, 209, 216, 232, 240, 243, 244
hybridisation. 48
Hyrax Hill. 49, 50, 52, 59, 196, 197, 209, 284
Ibn Batuta. 7, 178, 179, 186, 202, 211, 281
Ibn Majid. 173, 175, 177, 178, 187, 224, 225, 261, 264-267
ice-age. 23, 25, 27, 41, 52, 55, 119, 129
ice-cores. 16
Idrisi. 185
Inanda Dam. 193, 196, 212
India. . . . 1, 5, 7, 14, 15, 19, 43, 44, 50-52, 56, 60, 63-69, 103, 104, 107, 118-120, 125, 127-137, 139-143, 146-149,
151-155, 158, 161-164, 166, 170, 172-178, 182, 184, 186, 187, 189-191, 196-198, 202,
210, 211, 214, 219, 221, 225, 227-231, 238, 240, 242, 247, 248, 251, 253-272, 274, 282-
284
Indian. 1, 5, 14, 15, 19, 50, 52, 56, 60, 63-69, 103, 104, 119, 120, 127-131, 133, 134, 136, 137, 139-141, 143,
146-149, 151-155, 158, 161-164, 166, 170, 172-178, 184, 186, 187, 189-191, 196-198,
202, 210, 211, 214, 219, 221, 225, 227, 229-231, 240, 242, 251, 253-261, 264-272, 274,
282, 284
Indian Ocean. 1, 5, 14, 15, 19, 50, 56, 60, 65-67, 69, 103, 119, 127-131, 133, 134, 136, 137, 139-141, 143, 146-
149, 151-153, 155, 158, 161-164, 166, 170, 172-175, 177, 178, 186, 187, 189-191, 196-
198, 210, 211, 214, 219, 221, 225, 227, 229-231, 240, 242, 251, 253-261, 264-271, 274,
282, 284
Indonesia. 63-65, 108, 128, 140, 146-149, 162, 186, 282, 284
Indus. 35, 51, 96, 131, 134, 139, 151, 153, 287
industrial age. 140, 249
Inhaca Island. 176
Inhambane. 51, 174, 176, 188, 224-226, 236, 269
insects. 43, 73
Interlacustrine Zone. 45, 54-57, 60, 63-66, 68, 69, 161, 167, 168, 193, 199, 203, 208, 214, 215, 218, 230
iron. 19-21, 40, 42, 43, 45, 49-51, 53, 55, 57, 59, 66, 68, 69, 86, 94, 96, 98-102, 117, 120-123, 132, 133, 135,
136, 138, 141, 150, 154, 156, 158, 159, 161-163, 165, 168, 169, 177, 184, 186, 191-200,
202, 205, 208-210, 212, 214-216, 218, 219, 222, 223, 226-229, 231-233, 235-237, 242,
244, 249, 269, 280, 282-284, 287
Iron Age. . . 21, 40, 42, 43, 49-51, 55, 57, 59, 66, 68, 69, 86, 94, 96, 98-101, 117, 120-122, 135, 150, 156, 158, 159,
161, 162, 168, 169, 177, 192-200, 205, 208, 212, 214-216, 218, 219, 222, 223, 226, 229,
232, 233, 236, 237, 242, 249, 280, 282-284, 287
iron ore. 191, 195, 199
irrigation. 35, 36, 39, 41, 51-53, 58-60, 62, 137, 209, 279
Islam. 61, 139, 164-166, 170, 171, 174, 175, 187, 189, 190, 249, 254, 255
island-hopping. 128
Israel. 25, 135, 136, 143, 283
Israelites. 135, 142
Jacobson. 98, 280, 283
Jakubowski. 14, 283

Jane Goodall. 46
Jared Diamond. 108
Java. 281, 286
Jericho. 26, 38
Jews. 133, 137
Johanson. 283, 285
John Campbell. 243
Kaditshwene. 243
Kalahari. 73-76, 78, 79, 81, 82, 233, 281, 285
Kalambo Falls. 210
Kalundu pottery. 217, 229
Kamba. 123
Kaokoveld. 111, 282
Karnak. 132
Karoo. 79, 111
Kenya. 13, 14, 29, 40, 49, 50, 52-54, 56, 65, 69, 70, 89, 99, 111, 125, 159, 163, 168-170, 174, 177, 192, 196,
                                                                                    198, 203, 206-208, 212, 273, 282-285, 287
Kerio Valley. 53, 59
Khami. 21, 229, 239-241, 243
Khartoum. 67, 154
Khoekhoe. 71
Khoi. 27, 71, 81, 82, 86, 87, 89, 91-113, 117, 120, 121, 147, 194, 258, 269
KhoiKhoi. 87, 101, 102
Khoisan. 5, 44, 71, 72, 78, 81-89, 91, 95, 96, 106-110, 113, 115, 117, 120, 194, 195, 216, 218, 286
Kikuyu. 52
Kilimanjaro. 53, 65, 151
Kilwa. 9, 67, 149, 160, 161, 168, 172, 174-176, 178, 179, 181, 182, 186, 188, 192, 202, 207, 213, 221, 222, 228,
                                                                                                                                        270
Kinehan. 98, 283
King Solomon. 136, 137
Klasies River. 281
Kruger National Park. 233
Kunene River. 111
Kung. 48, 76, 77
Kwaaihoek. 258
Kwale. 66, 68-70, 156, 158, 159, 167-169, 192, 193, 197, 200, 212, 215, 216
Kwale pottery. 66, 68-70, 168, 169, 212, 216
KwaZulu-Natal. 14, 47, 51, 121, 191-195, 197-199, 201-205, 208-210, 212, 220, 222, 229, 243, 282, 287
Lake Chad. 43
Lake Malawi. 57, 228
Lake Tanganyika. 46, 57, 201, 210, 228
Lake Victoria. 53-55, 63, 64, 67, 68, 90, 124, 200, 205, 208-210, 228, 230, 273, 287
Lamu. 7, 10, 151, 155, 163, 167, 169, 172, 174, 177, 187, 222, 273
language. 20, 30, 50, 57, 71, 72, 76, 78, 87, 100, 102, 113, 138, 148, 153, 157, 158, 167, 168, 194, 209, 215,
                                                                                                                    217, 227, 229
Lascaux. 285
Late Iron Age. 21, 50, 55, 59, 98, 121, 177, 205, 222, 232, 233, 236, 242, 249, 282
Late Stone Age. 16, 19, 24, 30, 32, 36, 37, 39, 40, 44, 47-50, 53, 55, 56, 66, 69, 71, 76, 81, 82, 86, 88, 89, 92,
                                                           93, 96-98, 102, 105, 118, 120, 130, 145, 146, 156, 159, 160, 192, 193, 198, 214
Leakey. 49, 283, 285
Lebombo. 191
Levant. 24, 84, 141, 254, 279
Lewis-Williams. 33-35, 283
Limpopo Province. 241
Limpopo River. 162, 176, 185, 229, 231, 237
lipids. 280
literacy. 30
Livingstone. 152, 273, 274, 282, 283, 285
Lovelock. 283
Luanda. 257
Luangwa River. 228
Luxor. 132, 134
Lyall Watson. 48, 114, 224

Machemma. 241
MacMillan. 281, 283, 284
Madagascar. 50, 64, 108, 148, 149, 162, 186, 205, 271, 280
Mafia Island. 68, 149, 159, 160, 168, 174, 228, 280
Maggs. 14, 192, 193, 196, 284
Makgadikgadi. 282
Malabar Coast. 162, 163, 184, 256, 267, 270
Malacca. 163
malaria. 79, 281
Malawi. 50, 57, 65, 201, 204, 205, 208, 210, 212, 213, 217, 227, 228, 273
Malaysia. 50, 63, 163
Maldive Islands. 130
Mali. 200, 266
Malindi. 65, 163, 165, 174, 176, 207, 264-267, 273
Manda Island. 177
Mande. 125, 128, 154
Mapungubwe. 21, 229, 231, 232, 234-237, 239, 243, 244, 246, 282
Maputo. 21, 229
Marco Polo. 186, 283
Marib. 136
Mary Leakey. 49
Masai. 52, 122, 123
Mashatu Game Reserve. 232
Mashonaland. 239, 241, 279
Masudi. 184, 211, 234
Matopos. 79
Maya. 281
Mazel. 193, 194
McCarrison Society. 284
megalithic. 37, 41, 59, 130, 132, 146
Meroe. 125, 154
Mesopotamia. 20, 23, 35, 44, 47, 96, 131-133, 135, 141, 142, 151, 211
meteor. 142, 284
mfecane. 204-207, 241, 243
Mhlatuzana. 194
Michael Crawford. 14
Middle East. . . 17, 20, 26, 27, 31, 33, 36, 39, 43, 44, 48, 55, 60, 63, 65, 66, 118, 119, 137, 140, 173, 175, 189, 197, 284
Middle Stone Age. 24, 28, 86, 128, 150, 191, 194, 282
migration. 15, 28, 31, 36, 41, 51, 52, 55, 58, 61, 62, 65, 69, 82, 84, 89, 93, 97, 98, 100, 128, 129, 140, 146, 148, 161, 169, 192, 198-200, 203, 208, 217, 229, 237
Minoans. 132
Miocene. 280
mitochondrial. 85, 282
Mnarani. 203
Mngeni River. 196
Moçambique Island. 165, 175, 176, 178, 188, 222, 224, 272, 274, 277
Mogadiscio. 69, 176, 202
Mohenjo Daro. 134
Mombasa. 8, 12, 14, 66, 69, 151, 156, 168, 172, 174, 176, 187, 192, 197, 202, 203, 207, 270, 272, 273
Monica Wilson. 87, 101, 102
monkeys. 46, 139
Monomotapa. 225, 240, 241, 270
monsoon. 52, 56, 60, 66, 137, 140, 146, 149, 160, 164, 171, 174, 183, 198, 261-263, 282
Moors. 8, 152, 180-182, 250
Morais. 284
Morgan. 284
Moslem. 163, 252, 266, 268, 271
mosquitoes. 201
Mossel Bay. 103-105, 269
Motloutse River. 233, 246
Mouza. 155, 157, 158

Mozambique.... 1, 50, 51, 83, 84, 156, 159, 164, 174, 176, 178, 181, 182, 187, 188, 192, 193, 197, 202, 204, 221-223, 225, 231, 234, 252, 263, 269, 271, 279-281, 286
Mpumalanga... 192
mtwepe... 11, 183, 187
Muaconi... 197
Mungo... 124, 285
Mungo Park... 124, 285
music... 51, 73, 75, 104, 113, 244, 269
mutation... 16, 17, 25, 32, 33, 36, 287
mutations... 15, 28, 33, 119
Mzilikazi... 241
Mzonjani... 192, 284
Nabateans... 140, 142
Nairobi... 14, 208
naked ape... 284
Nakuru... 49, 52, 209
Nalatale... 241
Nama... 72, 87, 101, 102, 107, 110, 111, 197
Namaqualand... 102, 110
Namib Desert... 79, 257
Namibia... 72, 76, 77, 83, 84, 86, 87, 97-99, 102, 103, 107, 110, 111, 194, 257, 280, 281, 283, 285
Natal.... 1, 14, 47, 51, 121, 174, 191-195, 197-199, 201-205, 208-210, 212, 220, 222, 229, 243, 279, 281-284, 287
Natufians... 38
Naturics... 14, 283
navigation... 43, 44, 129, 140, 143, 146, 149, 151, 155, 173-178, 186, 189, 225, 251, 254, 258, 261, 263-266, 271, 284
Ndebele... 241, 243
Neanderthal... 285
Neccho... 143, 177
Negro... 20, 57, 72, 82, 83, 86, 94, 96, 101, 109, 120, 180, 216, 218, 269
Negroes... 19, 84, 86, 87, 96, 104, 105, 121, 167, 194, 255, 286
Nelson... 109, 136, 247-249, 284
Neolithic..... 26, 32-34, 36, 45, 57, 67, 68, 82, 90, 125, 129, 147, 159-161, 167-169, 193, 200, 218, 280, 283, 285
Nevali Cori... 26
Ngoni... 123, 205
Nguni... 21, 47, 102, 107, 109, 123, 126, 193-197, 204, 205, 207-209, 229, 239, 241, 243
Ngwenya Mountain... 191
Niger River... 41, 51, 68, 185
Nilotic... 50, 57, 60, 92, 111, 120, 124, 156, 159, 169, 170, 193, 208, 209, 218
Njemps... 53, 55, 58, 59
Nkope pottery... 217
nomadic... 5, 27-29, 31, 42, 47, 55, 71, 90, 111, 122, 124, 156, 195, 232
nomadism... 52, 96, 146
Nubia... 90, 99, 125, 134, 139, 143
nutrition... 14-16, 24, 27, 76, 122, 280, 281, 284, 285
nutritional driving force... 24
oceans... 128, 130, 132, 135, 143, 146, 149, 248, 253, 264
ochre... 191, 192, 282
Okavango... 76, 79, 101, 282
Olduvai... 283
Oliver... 38, 91, 92, 133, 285
Olorgasailie... 29, 30, 52
Oman... 11, 134, 166, 174, 177, 184, 185, 187, 266, 273, 279, 283
Ophir... 6, 136, 137, 139, 226, 255
Orce... 281
out of Africa... 128, 140, 284
Owens... 285
Pakefield... 285
Pakistan... 132, 151
Palestine... 6, 24, 26, 135, 140
Pangani River... 151
Paul Sinclair... 66, 67, 159, 188, 221, 222
Pemba Island... 155

Pemba River. 66, 68, 212
Periplus. 7, 153-156, 159, 161, 164, 165, 170, 173, 198, 225, 280, 282, 286
Periplus of the Erythraean Sea. 153, 161, 198, 282, 286
Persian. 130, 133-135, 139, 142, 143, 145, 150, 151, 153, 164, 170, 174, 175, 202, 222, 227, 254, 272
Persian Gulf. 130, 133-135, 139, 142, 150, 151, 153, 164, 254, 272
Phalaborwa. 232
Philistines. 135, 136, 151
Phillip Tobias. 14, 79, 83, 195
Phoenicians. 133, 136, 142-144, 147, 280
Pietermaritzburg. 193, 194, 212
Pleistocene. 279, 281, 285, 287
Pliny. 152, 155
Pole Star. 175, 251
Polynesia. 130, 146
Polynesians. 130, 146
Portuguese. 1, 19, 63, 78, 97, 103, 120, 143, 149, 150, 163-166, 169, 170, 172, 173, 175-177, 179, 181, 182,
 184, 188-190, 204-207, 220-222, 224, 225, 228, 229, 236, 239-241, 248-251, 253-256,
 258-275, 277, 279, 280, 284
pottery. 26, 38-40, 43, 50, 51, 66-70, 89, 92-94, 97-100, 102, 103, 105, 131, 156, 158-161, 168-170, 192-194,
 196, 197, 199-201, 210, 212, 214-219, 222, 224, 229, 280, 283
Prester John. 255, 258
primates. 46, 114
Ptolemy. 152, 155, 159, 160, 165
Puebla. 128
Punt. 137, 150, 226
pyramids. 23, 37, 54, 284
Quelimane. 222
race. 71, 72, 81-85, 87, 94, 100, 105-107, 120, 136, 168, 186, 202, 244, 247, 280
radiation. 15, 16, 35, 36, 48, 119
rafts. 128, 130, 251
rainforest. 55, 76, 92, 123, 189, 198, 228, 254
Red Sea. 43, 44, 64, 125, 128, 132-140, 142, 143, 149-151, 153-155, 165, 173, 174, 177, 189, 253, 282, 284
Redfield. 14, 285
Rhapta. 69, 155-160, 165, 225, 228
Richard Wilding. 14, 89, 156, 167, 192, 202
rock-art. 14, 39, 118, 191
Romans. 132, 140, 147, 162, 164, 165, 167, 268, 280
Rovuma. 65
Rozwi. 239-241
Ruaha. 228
Rufiji. 66, 67, 69, 159, 160, 165, 225, 228
Saba. 9, 60, 137, 140, 158
Sagres. 251, 252
Sahara. . . . 1, 5, 14, 17, 19, 21, 36, 38-46, 50, 55, 60, 61, 63, 66, 68, 74, 81, 83, 85, 86, 88, 102, 105, 111, 118-125,
 135, 139, 145-147, 150, 161, 167, 170, 194, 211, 216, 226, 247-249, 251-254, 272, 280,
 283, 286
Sahel. 24, 41, 42, 44, 45, 55, 59, 60, 66, 68, 119, 122, 125, 135, 161, 200, 206, 251
Salalah. 154
Samburu. 111
San. . . 7, 42, 46, 48, 55, 57, 71, 74-76, 78-82, 86, 87, 95, 96, 100, 103-106, 109, 113, 146, 150, 178, 192, 194, 195,
 281, 283
sanga. 55, 119-121, 286
San-Bushman. 81
São Salvador. 257
savannah. 23, 27, 38, 41, 42, 54-57, 65, 66, 69, 76, 79, 81-83, 86, 92, 118, 120, 198, 201, 209, 216-218, 284
Save River. 174, 176, 222, 225, 235, 286
seafood. 15, 16, 216
seashore. 1, 4, 14, 15, 18, 19, 35, 56, 84, 130, 193, 279
Seashore Hypothesis. 15, 130
seatraders. 3, 5, 19, 29, 63, 65, 127, 142, 146, 149, 153, 155, 156, 158, 160, 164, 173, 177, 183, 186, 196, 197,
 199, 203, 219, 221, 226, 227, 230-232, 254, 265, 267-269
Senegal. 253
Shaka. 95, 121, 123, 204, 205, 209, 234, 243

Sheba... 136, 137
sheep... 23, 27, 38, 42, 63, 66, 71, 72, 88, 92, 93, 97-101, 104, 105, 117, 118, 122, 179, 194-196, 218, 244
shellfish... 280
Shona... 207, 229, 234, 235, 237-239, 241, 242, 279
Sierra Leone... 145, 253
Silver Leaves... 192
Simba... 207
Sinai... 135
Sirikwa Holes... 59, 196, 209
slavery... 29, 47, 48, 106
slave-trade... 107
Sofala... 8, 9, 102, 159, 164, 176, 178, 182, 184, 186, 188, 189, 211, 220-222, 224, 225, 228, 235, 236, 240, 258, 262-264, 272
Sotho... 101, 102, 107, 123, 204, 208, 229, 243, 285
Sotho-Tswana... 229
South Africa... 1, 13, 14, 27, 42, 43, 46, 50, 51, 70-73, 75, 76, 86, 87, 92, 101, 102, 106-111, 113, 115, 120, 123, 126, 142, 150, 166, 177, 188, 200, 201, 203, 204, 206, 210, 214, 217, 231, 233, 237, 241, 243, 249, 269, 274, 279-287
southern Africa... 5, 19, 21, 34, 43, 48, 54, 55, 57, 60, 61, 71, 72, 79, 82, 84, 91-94, 97-102, 107, 110-112, 115, 117, 120, 122-125, 147, 159, 162, 165, 167, 191-194, 200-202, 205, 207-210, 212, 214-219, 221-224, 226-232, 234, 236, 237, 240, 242, 254, 262, 273, 280, 281, 286, 287
St. Helena Bay... 104, 263
Sterkfontein... 280
Stone Age... 16, 19, 24, 28-30, 32, 36, 37, 39, 40, 44, 47-50, 53-56, 66, 69, 71, 73, 74, 76, 81, 82, 86, 88, 89, 92, 93, 96-98, 102, 105, 106, 118, 120, 128-130, 145, 146, 150, 156, 159, 160, 191-194, 196, 198, 214, 281, 282, 285
Stringer, Chris... 286
sub-Sahara Africa... 55, 68, 249
Sudan... 42, 43, 51, 59, 60, 68, 120, 125, 154, 156, 200, 227, 254, 279
Suez... 36, 134
Sumerians... 133
Sunda... 104, 121, 142, 148, 280, 283, 284, 286
supernova... 16, 286
supernovae... 16
Swahili... 5, 20, 50, 58, 153, 159-161, 164-173, 177, 179, 182, 183, 186-188, 202, 203, 211, 221, 222, 224, 225, 228, 231, 236, 272-274, 280, 282, 287
Swaziland... 191, 204, 224
swimming... 130
Tana River... 159, 160, 215
Tanganyika-Malawi gap... 217, 228
Terra da Boa Gente... 269
Tete... 188, 207, 228
Thatta... 151
Thebes... 132, 134
Thukela River... 193
Thulamela... 185, 233-235, 241, 244
Tibbetts... 8, 149, 173, 175, 176, 225, 265, 267
Tim Maggs... 14, 192, 193, 196
Timor... 128
Tiwi Beach... 12
Tobias... 14, 79, 83, 195, 286
Tom Huffman... 14, 21, 40, 201, 208, 210, 214, 229, 233, 237, 242, 243
trade... 11, 19-21, 29, 31, 38, 40, 43, 50, 51, 60, 63, 65-67, 104, 107, 122, 125, 134-137, 139-142, 146-149, 151, 153-157, 160-162, 164-167, 170-172, 174-176, 182, 186-190, 197, 202, 204, 209-211, 221-233, 236-241, 248-255, 258, 259, 261, 267-274, 282, 287
trading... 1, 5, 19-23, 28-30, 39, 42, 44, 52, 58, 60, 64, 65, 68, 69, 101, 108, 112, 119, 123, 127, 134, 136, 138-143, 145, 147, 150-156, 158, 160-165, 170-175, 182, 184, 188, 189, 198, 202-208, 210, 211, 214, 219, 221-231, 233, 236-238, 240-243, 248, 250, 251, 253-256, 258-260, 265, 268-270, 272-274, 286
Transkei... 193, 209
Transvaal... 107, 192, 237, 243, 244
Tristão... 253
tsetse-fly... 101, 226, 231

Tswana.. 42, 204, 205, 208, 229, 234, 243, 244
Tugen hills.. 53
Turkana... 40, 41, 52
Urewe pottery.. 66, 68, 217, 218
van der Post.. 75, 81
Versailles... 247
Victoria................. 53-55, 63-65, 67, 68, 89, 90, 124, 188, 200, 205, 208-210, 228, 230, 273-275, 287
Wadi Hammamat.. 133
Watson... 48, 114, 224, 287
Wendt.. 287
White Nile.. 124, 284
Wilding.. 14, 89, 156, 167, 174, 192, 202, 203, 287
Witwatersrand.. 14, 142, 214, 280, 282
XaiXai... 192, 197
Xhosa.. 108, 109, 193, 204
Yemen................. 60, 120, 125, 134, 136, 137, 140, 154, 155, 158, 163, 170, 174, 198, 283, 287
Zambezi River.. 101, 188
Zambia... 43, 57, 92, 198, 201, 205, 212, 236, 283
Zanj.. 7, 8, 184, 202, 262
Zanzibar...... 1, 6-9, 14, 66-69, 92, 141, 148, 149, 153-155, 159, 166, 168, 171, 172, 174, 176, 178, 182, 183, 186,
                                                                                    202, 212, 271, 273
zebu... 119
Zheng He... 162, 163, 268, 283
Zimbabwe... 5, 21, 43, 50, 59, 60, 62, 72, 92, 99, 160, 162, 164, 171, 182, 201, 204, 207, 210-212, 221-225, 227-
                                                                245, 270, 271, 279, 281, 282, 286
Zulu....... 13, 14, 47, 51, 121, 123, 191-195, 197-199, 201-205, 208-210, 212, 220, 222, 229, 243, 282, 284, 287
Zululand... 95, 281
Zumbo.. 228

African Insight, 2008
www.sondela.co.uk

Made in the USA
San Bernardino, CA
28 August 2014